Introduction to the Laplace Transform

MATHEMATICAL CONCEPTS AND METHODS IN SCIENCE AND ENGINEERING

Series Editor: **Angelo Miele**
Mechanical Engineering and Mathematical Sciences
Rice University

Introduction to the Laplace Transform

Peter K. F. Kuhfittig

Milwaukee School of Engineering
Milwaukee, Wisconsin

PLENUM PRESS · NEW YORK AND LONDON

Library of Congress Cataloging in Publication Data

Kuhfittig, Peter, K. F.
    Introduction to the Laplace transform.

    (Mathematical concepts and methods in science and engineering)
    Includes index.
    1. Laplace transformation. I. Title.
    QA432.K78   515'.723   77-29017
    ISBN 0-306-31060-0

First Printing – April 1978
Second Printing – July 1980

© 1978 Plenum Press, New York
A Division of Plenum Publishing Corporation
227 West 17th Street, New York, N.Y. 10011

Printed in the United States of America

To Paul and Annemarie

Preface

The purpose of this book is to give an introduction to the Laplace transform on the undergraduate level. The material is drawn from notes for a course taught by the author at the Milwaukee School of Engineering. Based on classroom experience, an attempt has been made to (1) keep the proofs short, (2) introduce applications as soon as possible, (3) concentrate on problems that are difficult to handle by the older classical methods, and (4) emphasize periodic phenomena.

To make it possible to offer the course early in the curriculum (after differential equations), no knowledge of complex variable theory is assumed. However, since a thorough study of Laplace transforms requires at least the rudiments of this theory, Chapter 3 includes a brief sketch of complex variables, with many of the details presented in Appendix A. This plan permits an introduction of the complex inversion formula, followed by additional applications.

The author has found that a course taught three hours a week for a quarter can be based on the material in Chapters 1, 2, and 5 and the first three sections of Chapter 7. If additional time is available (e.g., four quarter-hours or three semester-hours), the whole book can be covered easily.

The author is indebted to the students at the Milwaukee School of Engineering for their many helpful comments and criticisms.

Milwaukee, Wisconsin   P. K. F. Kuhfittig

Contents

1

Basic Properties and Applications

1.1. Introduction

Let $f(t)$ be a function defined on a finite or infinite interval $a < t < b$, and let $k(s, t)$ be a prescribed function of the variable t and parameter s. Then

$$G(s) = \int_a^b k(s, t) f(t)\, dt$$

is called an *integral transform* of f. One of the simplest and most useful of these transforms is the *Laplace transform*,

$$L\{f(t)\} = F(s)$$

$$= \int_0^\infty e^{-st} f(t)\, dt,$$

where s is a complex variable.

Laplace transforms may be used to solve differential equations. To see how, let us find an expression for $L\{f'(t)\}$. By formal in-

1

tegration by parts

$$L\{f'(t)\} = \int_0^\infty e^{-st} f'(t)\, dt$$

$$= \lim_{b \to \infty} \int_0^b e^{-st} f'(t)\, dt$$

$$= \lim_{b \to \infty} \left[f(t)e^{-st} \Big|_0^b + s \int_0^b f(t)e^{-st}\, dt \right]$$

$$= \lim_{b \to \infty} \left[f(b)e^{-sb} - f(0) \right] + sL\{f(t)\}$$

$$= sL\{f(t)\} - f(0),$$

assuming that $\lim_{b \to \infty} f(b)e^{-sb} = 0$. We see, then, that the Laplace transform of f' has been expressed in terms of the transform of f itself.

Now consider the differential equation

$$f'(t) - f(t) = 0,$$

satisfying the initial condition $f(0) = 1$, and let us find the Laplace transform of both sides of the equation. Clearly, $L\{0\} = 0$, so that

$$sL\{f(t)\} - f(0) - L\{f(t)\} = 0.$$

But this is just a simple *algebraic* equation! Solving for $L\{f(t)\}$ and using the condition $f(0) = 1$, we have

$$L\{f(t)\} = \frac{1}{s-1}.$$

We shall see in Section 1.3 that $L\{e^t\} = 1/(s-1)$. Consequently,

$$f(t) = e^t$$

must be the solution of the equation.

The procedure for solving differential equations by Laplace transforms may be summarized as follows:

(a) Find the Laplace transform of both sides of the differential equation.
(b) Solve the resulting algebraic equation for $L\{f(t)\}$.
(c) Find the *inverse transform*.

It is clear from the example that the method does more than just convert differential equations into algebraic equations: Since the initial conditions are incorporated in the transformed equation, no arbitrary constants appear in the solution. Other, even greater, advantages of the Laplace transform method will become apparent later.

Simple though the above procedure may appear, to use this method effectively, we are going to need the transforms of a number of specific functions as well as a knowledge of the more important properties of the Laplace transform.

1.2. The Variable s

We noted in Section 1.1 that

$$F(s) = \int_0^\infty e^{-st} f(t)\, dt \qquad (1.1)$$

is a function of a complex variable. In fact, we shall see in Chapter 4 that $F(s)$ is an analytic (differentiable) function and that the integration in equation (1.1) can be performed as if s were a real variable. For example, if $f(t) = 1$,

$$\int_0^\infty e^{-st}\, dt = \frac{1}{s},$$

for $s > 0$, s real. If s is complex, the condition $s > 0$ is replaced by $\mathrm{Re}(s) > 0$ (real part of s is greater than 0). In most cases we need the condition $s > \alpha$; in Chapter 4 this becomes $\text{Re}(s) > \alpha$, from which it will follow that $F(s)$ is analytic in the half-plane $\text{Re}(s) > \alpha$. Now in the first two chapters no explicit use is ever made of the number α in solving problems, the existence of such a number being enough —and the existence is important only in the sense that we need to ensure the convergence of the improper integrals involved. It is therefore convenient to take s to be real for now and to postpone the discussion of complex s until later chapters, where the fact that $F(s)$ is analytic in a half-plane will play an important role.

1.3. Laplace Transforms of Some Special Functions

Example 1.1. Suppose we find the transform of the function $f(t) = 1$ for $t > 0$ by using the definition. Then

$$L\{1\} = \int_0^\infty 1 \cdot e^{-st}\, dt$$

$$= \lim_{b \to \infty} \int_0^b e^{-st}\, dt$$

$$= \lim_{b \to \infty} \left(-\frac{1}{s} e^{-st} \Big|_0^b \right)$$

$$= \lim_{b \to \infty} \left(-\frac{1}{s} e^{-sb} + \frac{1}{s} \right).$$

If $s > 0$, we have

$$L\{1\} = \frac{1}{s} \qquad (s > 0). \tag{1.2}$$

We have already noted that the result is also valid if s is a complex variable whose real part is positive.

Since the limits of integration are 0 and ∞, there is no apparent need to consider the behavior of $f(t)$ for negative t. Later, however, we shall find it convenient to define $f(t)$ as 0 for all negative values.

Example 1.2. For $t > 0$

$$L\{e^{at}\} = \int_0^\infty e^{at}e^{-st}\,dt$$

$$= \lim_{b \to \infty} \int_0^b e^{-(s-a)t}\,dt$$

$$= \lim_{b \to \infty} \left[\frac{e^{-(s-a)t}}{-(s-a)} \Big|_0^b \right]$$

$$= \lim_{b \to \infty} \left[\frac{e^{-(s-a)b}}{-(s-a)} + \frac{1}{s-a} \right]$$

$$= \frac{1}{s-a},$$

provided that $s > a$ [or for complex s, $\mathrm{Re}(s) > a$]. Thus

$$L\{e^{at}\} = \frac{1}{s-a} \qquad (s > a). \tag{1.3}$$

Example 1.3. As another example, let us consider $f(t) = t^n$, $t > 0$, and n a nonnegative integer. Integrating by parts,

$$L\{t^n\} = \int_0^\infty t^n e^{-st}\,dt$$

$$= -\frac{1}{s} t^n e^{-st} \Big|_0^\infty + \frac{n}{s} \int_0^\infty t^{n-1} e^{-st}\,dt$$

$$= \frac{n}{s} L\{t^{n-1}\}$$

by L'Hospital's rule. Similarly,

$$L\{t^{n-1}\} = \frac{n-1}{s} L\{t^{n-2}\},$$

and, after n steps,

$$L\{t^0\} = L\{1\} = \frac{1}{s} \qquad (s > 0)$$

by equation (1.2). Combining these results,

$$L\{t^n\} = \frac{n(n-1)(n-2)\cdots 2 \cdot 1}{s^{n+1}}$$

$$= \frac{n!}{s^{n+1}} \qquad (s > 0). \tag{1.4}$$

This transform reduces to equation (1.2) if $n = 0$.

Example 1.4. The transform of $f(t) = t^n$ can be generalized to t^α, $t > 0$, where α is any real number greater than -1. Making the change of variable $x = st$,

$$L\{t^\alpha\} = \int_0^\infty t^\alpha e^{-st}\, dt$$

$$= \frac{1}{s^{\alpha+1}} \int_0^\infty x^\alpha e^{-x}\, dx \qquad (\alpha > -1,\ s > 0).$$

The integral $\int_0^\infty x^\alpha e^{-x}\, dx$ is known as *Euler's gamma function* and is denoted by $\Gamma(\alpha + 1)$. Even though the integrand has no elementary antiderivative, the integral exists for $\alpha > -1$ (more precisely, for α not equal to a negative integer), and tables have been tabulated for different values of α. Thus

$$L\{t^\alpha\} = \frac{\Gamma(\alpha + 1)}{s^{\alpha+1}} \qquad (\alpha > -1,\ s > 0). \tag{1.5}$$

We can see from equation (1.4) that

$$\Gamma(n + 1) = n!$$

if n is a nonnegative integer, so that the gamma function may be viewed as a generalization of the factorial.

An interesting special case arises by taking $\alpha = -\frac{1}{2}$. From the theory of probability it is known that

$$\frac{2}{\pi^{1/2}} \int_0^\infty e^{-x^2}\, dx = 1.$$

So by letting $x = (st)^{1/2}$, we obtain

$$\int_0^\infty t^{-1/2} e^{-st}\, dt = \int_0^\infty \frac{s^{1/2}}{x} e^{-x^2} \frac{2x\, dx}{s}$$

$$= \frac{2}{s^{1/2}} \int_0^\infty e^{-x^2}\, dx$$

$$= \left(\frac{\pi}{s}\right)^{1/2} \qquad (s > 0),$$

that is,

$$L\{t^{-1/2}\} = \left(\frac{\pi}{s}\right)^{1/2} \qquad (s > 0). \tag{1.6}$$

Before continuing with our next example, let us note that for any constants a and b and any two integrable functions f and g

$$\int [af(x) + bg(x)]\, dx = a \int f(x)\, dx + b \int g(x)\, dx;$$

hence

$$L\{af(t) + bg(t)\} = aL\{f(t)\} + bL\{g(t)\}, \tag{1.7}$$

if the transforms exist. Since it possesses property (1.7), the Laplace transform is said to be *linear*.

Example 1.5

$$L\{\cosh at\} = L\left\{\frac{e^{at} + e^{-at}}{2}\right\}$$

$$= \frac{1}{2}\left(L\{e^{at}\} + L\{e^{-at}\}\right)$$

$$= \frac{1}{2}\left(\frac{1}{s-a} + \frac{1}{s+a}\right)$$

by equations (1.3) and (1.7). Thus

$$L\{\cosh at\} = \frac{s}{s^2 - a^2} \qquad (s > |a|). \qquad (1.8)$$

Example 1.6

$$L\left\{\sum_{k=0}^{n} a_k t^k\right\} = \sum_{k=0}^{n} a_k L\{t^k\} = \sum_{k=0}^{n} a_k \frac{\Gamma(k+1)}{s^{k+1}}$$

$$= \frac{a_0}{s} + \frac{a_1}{s^2} + \frac{2! \, a_2}{s^3} + \frac{3! \, a_3}{s^4} + \cdots + \frac{n! \, a_n}{s^{n+1}}$$

$$(s > 0), \qquad (1.9)$$

since k is a nonnegative integer. This is the formula for the Laplace transform of a polynomial of degree n.

The linearity property can be applied to functions having a power series expansion. Suppose

$$f(t) = a_0 + a_1 t + a_2 t^2 + \cdots = \sum_{n=0}^{\infty} a_n t^n;$$

then

$$F(s) = \frac{a_0}{s} + \frac{a_1}{s^2} + \frac{2! \, a_2}{s^3} + \frac{3! \, a_3}{s^4} + \cdots = \sum_{n=0}^{\infty} \frac{n! \, a_n}{s^{n+1}} \qquad (s > 0),$$

provided, of course, that the series involved converge.

Example 1.7. If

$$f(t) = e^t = \sum_{n=0}^{\infty} \frac{t^n}{n!},$$

then

$$F(s) = \sum_{n=0}^{\infty} \left(\frac{1}{s}\right)^{n+1} = \frac{1}{s-1},$$

which converges for $s > 1$, in agreement with equation (1.3). Here we have used the formula for the sum of a geometric series:

$$\sum_{n=0}^{\infty} ar^n = \frac{a}{1-r} \qquad (|r| < 1).$$

For completeness we shall state without proof that if

$$L\{f(t)\} = \sum_{n=0}^{\infty} a_n s^{-n-1}$$

converges for $s > b$, then $\sum_{n=0}^{\infty} (a_n t^n / n!)$ also converges, and

$$f(t) = \sum_{n=0}^{\infty} \frac{a_n t^n}{n!} \qquad (t > 0);$$

the function $f(t)$ is called the *inverse transform*.

Transforms of other specific functions can be obtained by making use of various properties of the transform to be discussed in later sections. (A short table of Laplace transforms can be found in Appendix B.)

Exercises

1.1. Find the Laplace transform of each of the following functions:

(a) $L\{2 + 3t\}$
(b) $L\{3t + 5t^3\}$
(c) $L\{5e^t\}$
(d) $L\{4e^{-3t}\}$
(e) $L\{\cosh 5t\}$

1.2. Show that

(a) $L\{\cos at\} = \dfrac{s}{s^2 + a^2} \qquad (s > 0)$

(b) $L\{\sin at\} = \dfrac{a}{s^2 + a^2} \qquad (s > 0)$

by using the definition.

1.3. Show that

$$L\{\sinh at\} = \frac{a}{s^2 - a^2} \qquad (s > |a|),$$

where $\sinh at = \frac{1}{2}(e^{at} - e^{-at})$.

1.4. The following relationships will be used in Chapter 3. Given the *Euler identity*

$$e^{iat} = \cos at + i \sin at,$$

one can easily check that

$$\sin at = \frac{e^{iat} - e^{-iat}}{2i}$$

and

$$\cos at = \frac{e^{iat} + e^{-iat}}{2},$$

where $i = \sqrt{-1}$. Use the relationships formally to obtain the transforms of the functions in Exercise 1.2.

1.5. Find
  (a) $L\{\sinh at\}$ by use of infinite series
  (b) $f(t)$ if $F(s) = s/(s^2 - a^2)$ by use of infinite series

1.6. Expand the following functions in a Maclaurin series, and find the transforms termwise:
  (a) $\sin t^2$
  (b) $\cos t^{1/2}$

1.7. Find $L\{J_0(t)\}$, where

$$J_0(t) = 1 - \frac{t^2}{2^2} + \frac{t^4}{2^2 \cdot 4^2} - \frac{t^6}{2^2 \cdot 4^2 \cdot 6^2} + \cdots.$$

$J_0(t)$ is called the *Bessel function* of order 0.

1.4. Some Basic Properties of the Laplace Transform

Theorem 1.1. First Translation Theorem. If $F(s) = L\{f(t)\}$ exists for $s > 0$, then for any constant a

$$L\{e^{at}f(t)\} = F(s - a) \qquad (s > a).$$

Proof

$$F(s - a) = \int_0^\infty e^{-(s-a)t} f(t) \, dt$$

$$= \int_0^\infty e^{-st} [e^{at} f(t)] \, dt$$

$$= L\{e^{at} f(t)\}.$$

Theorem 1.2. Change of Scale. If $F(s) = L\{f(t)\}$ exists for $s > s_0$, then $L\{f(at)\} = (1/a)F(s/a)$, $a > 0$, $s > as_0$.

Proof

$$F\left(\frac{s}{a}\right) = \int_0^\infty e^{-s(t/a)} f(t) \, dt.$$

Letting $t = ax$, we get

$$F\left(\frac{s}{a}\right) = \int_0^\infty e^{-sx} f(ax) a \, dx$$

$$= aL\{f(at)\}.$$

Definition 1.1. A function f is said to have a *finite discontinuity* at a if $f(a+) = \lim_{t \to a+} f(t)$ and $f(a-) = \lim_{t \to a-} f(t)$ exist. By $t \to a^+$ we mean that t approaches a from the right, through values larger than a. Similarly, $t \to a^-$ means that t approaches a from the left [Figure 1.1(a)].

Definition 1.2. A function f is said to be *sectionally continuous* for $t \geq 0$ if it has at most a finite number of finite discontinuities in the interval $0 \leq t \leq b$ for $b > 0$.

Example 1.8. The function

$$u(t - t_0) = \begin{cases} 1, & t > t_0 \\ 0, & t < t_0 \end{cases} \tag{1.10}$$

has a *jump* of 1 unit at t_0 [Figure 1.1(b)]. This function is called *Heaviside's unit function* and will play a key role in our later work. It is to be noted that the definition of a finite discontinuity of f at a says nothing about the value of f at a but only that the left and right limits must exist—$f(a)$ may or may not exist. Consequently, the function [again denoted by $u(t - t_0)$]

$$u(t - t_0) = \begin{cases} 1, & t \geq t_0 \\ 0, & t < t_0 \end{cases} \qquad (1.11)$$

is also sectionally continuous and may be more convenient than the function (1.10) in our later applications.

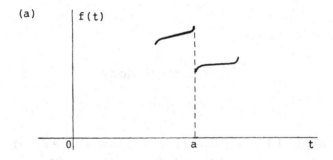

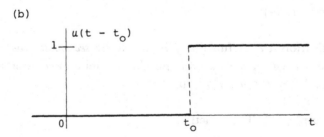

FIGURE 1.1. (a) A function with a finite discontinuity at a; (b) Heaviside's unit function.

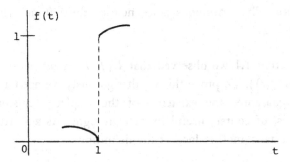

FIGURE 1.2. A function with a jump discontinuity at 1, where $f(1+) = 1$ and $f(1-) = 0$.

Example 1.9

$$\alpha(t) = \begin{cases} 1, & t \geq 0 \\ 0, & t < 0 \end{cases}$$

has a jump of 1 unit at $t = 0$.

Example 1.10. If

$$f(t) = \frac{1}{1 + 2^{-1/(t-1)}},$$

then

$$f(1+) = \lim_{t \to 1^+} \frac{1}{1 + 2^{-1/(t-1)}} = 1$$

and

$$f(1-) = \lim_{t \to 1^-} \frac{1}{1 + 2^{-1/(t-1)}} = 0,$$

showing that the function has a jump of 1 unit at 1. In this case $f(1)$ is not defined, yet f is sectionally continuous (Figure 1.2).

Example 1.11

$$f(t) = \frac{1}{t - 2}$$

is not sectionally continuous since neither the left nor right limits exist at $x = 2$.

In Section 1.1 we observed that $L\{f'(t)\}$ may be expressed in terms of $L\{f(t)\}$. To prove this result rigorously we need a condition that will guarantee the existence of the Laplace transform. This condition is, of course, useful in its own right. As a starting point, suppose we take another look at the definition

$$L\{f(t)\} = \int_0^\infty e^{-st} f(t)\, dt.$$

Since the integral is improper, it is intuitively obvious that e^{-st} has to dominate $f(t)$ in the sense that $f(t)$ must not grow faster than a function of exponential type, suggesting the following definition.

Definition 1.3. A function $f(t)$ is said to be of exponential order $e^{\alpha t}$ if there exist real constants $M > 0$ and $N > 0$ such that for all $t > N$

$$| e^{-\alpha t} f(t) | < M \qquad \text{or} \qquad | f(t) | < M e^{\alpha t}.$$

Example 1.12. Both $\sin at$ and $\cos at$ are of exponential order e^t since

$$| e^{-t} \sin at | < 1 \qquad \text{for } t > 1.$$

Here we have taken α, M, and N equal to 1. (Actually, the condition is satisfied for $t > 0$.)

Example 1.13. e^{3t} is of exponential order e^{at} for any $a > 3$, since

$$| e^{-at} e^{3t} | = e^{(3-a)t} < 1 \qquad \text{for } t > 1.$$

Clearly, e^{t^2} is not of exponential order.

Theorem 1.3. If $f(t)$ is of exponential order $e^{\alpha t}$ for some α and sectionally continuous for $t \geq 0$, its Laplace transform exists for $s > \alpha$.

Proof

$$L\{f(t)\} = \int_0^\infty e^{-st} f(t)\, dt$$

$$= \int_0^N e^{-st} f(t)\, dt + \int_N^\infty e^{-st} f(t)\, dt.$$

On the interval $[0, N]$, f has a finite number of jump discontinuities, so that the integral can be written as a finite sum of integrals in each of which the integrand is continuous—and continuity is a sufficient condition for integrability. For the second integral we use the fact that f is of exponential order. To see this we write

$$\left| \int_N^\infty e^{-st} f(t)\, dt \right| \leq \int_N^\infty |e^{-st} f(t)|\, dt$$

$$\leq \int_N^\infty e^{-st} M e^{\alpha t}\, dt = \frac{M}{s - \alpha}\, e^{-(s-\alpha)N}.$$

Consequently, the Laplace transform exists for $s > \alpha$. (It also follows that if $s \geq b > \alpha$, then for given $\varepsilon > 0$ and N sufficiently large, $|\int_a^\infty e^{-st} f(t)\, dt| < \varepsilon$ whenever $a > N$; N does not depend on s. These facts are needed in the proof of Theorem 1.6.)

It should be pointed out that the conditions of the theorem are sufficient but not necessary: The conditions are enough to guarantee the existence of the transform, but a function not satisfying these conditions may have a transform anyway. Fortunately, they provide an adequate check for most functions normally encountered.

We are now ready to state the main theorem of this section.

Theorem 1.4. Derivative Theorem. If $f(t)$ is continuous for $t > 0$ and of exponential order $e^{\alpha t}$ and if $f'(t)$ is sectionally continuous for $t \geq 0$, then

$$L\{f'(t)\} = sL\{f(t)\} - f(0+) \qquad (s > \alpha). \qquad (1.12)$$

Proof. Using integration by parts, we get, for $\varepsilon > 0$,

$$L\{f'(t)\} = \int_0^\infty e^{-st} f'(t)\, dt$$

$$= \lim_{\substack{b \to \infty \\ \varepsilon \to 0}} \int_\varepsilon^b e^{-st} f'(t)\, dt$$

$$= \lim_{\substack{b \to \infty \\ \varepsilon \to 0}} \left[e^{-st} f(t) \Big|_\varepsilon^b + s \int_\varepsilon^b e^{-st} f(t)\, dt \right]$$

$$= \lim_{\substack{b \to \infty \\ \varepsilon \to 0}} \left[e^{-sb} f(b) - e^{-s\varepsilon} f(\varepsilon) + s \int_\varepsilon^b e^{-st} f(t)\, dt \right]$$

$$= -f(0+) + s \int_0^\infty e^{-st} f(t)\, dt$$

$$= sL\{f(t)\} - f(0+).$$

We have used the fact that $\lim_{b \to \infty} e^{-sb} f(b) = 0$ for $s > \alpha$. To see why, we first observe that $| e^{-\alpha t} f(t) | < M$ for $t > N$, since $f(t)$ is of exponential order $e^{\alpha t}$. Now if $s > \alpha$, then $s = \alpha + \eta$ for some positive number η and

$$| e^{-st} f(t) | = | e^{-(\alpha+\eta)t} f(t) |$$

$$= | e^{-\alpha t} e^{-\eta t} f(t) |$$

$$= | e^{-\alpha t} f(t) | \, e^{-\eta t}$$

$$< M e^{-\eta t}$$

for $t > N$, but $\lim_{t \to \infty} M e^{-\eta t} = 0$.

Remark 1.1. If f is continuous for $t \geq 0$, $f(0+) = f(0)$, and equation (1.12) reduces to

$$L\{f'(t)\} = sL\{f(t)\} - f(0) \qquad (s > \alpha). \qquad (1.13)$$

Theorem 1.5. If f satisfies the conditions of Theorem 1.4 except for a finite discontinuity at $t = t_1$, then

$$L\{f'(t)\} = sL\{f(t)\} - f(0+) - e^{-t_1 s}[f(t_1+) - f(t_1-)] \qquad (s > \alpha).$$

Proof

$$L\{f'(t)\} = \lim_{b \to \infty} \int_0^b e^{-st} f'(t)\, dt$$

$$= \lim_{b \to \infty} \left[e^{-st} f(t) \Big|_0^b + s \int_0^b e^{-st} f(t)\, dt \right]$$

$$= \lim_{b \to \infty} \left[e^{-st} f(t) \Big|_0^{t_1-} + e^{-st} f(t) \Big|_{t_1+}^b + s \int_0^b e^{-st} f(t)\, dt \right]$$

$$= \lim_{b \to \infty} \left[e^{-st_1} f(t_1-) - f(0+) + e^{-sb} f(b) - e^{-st_1} f(t_1+) \right.$$

$$\left. + s \int_0^b e^{-st} f(t)\, dt \right]$$

$$= sL\{f(t)\} - f(0+) - e^{-st_1}[f(t_1+) - f(t_1-)].$$

Clearly, $|f(t_1+) - f(t_1-)|$ is the magnitude of the jump at $t = t_1$.

The theorem can be extended to any finite number of finite discontinuities.

Corollary 1.1. If $f(t)$ is sectionally continuous for $t \geq 0$ with at most a finite number of discontinuities and of exponential order $e^{\alpha t}$ and if $f'(t)$ is sectionally continuous, then

$$L\{f'(t)\} = sL\{f(t)\} - f(0+) - \sum_{i=1}^{n} e^{-st_i}[f(t_i+) - f(t_i-)] \qquad (s > \alpha),$$

where $t_1, t_2, \ldots, t_n$ are the positive abscissas of the points of discontinuity.

If, under suitable conditions, we apply Theorem 1.4 to f'', we get

$$L\{f''(t)\} = sL\{f'(t)\} - f'(0+)$$
$$= s[sL\{f(t)\} - f(0+)] - f'(0+),$$

or

$$L\{f''(t)\} = s^2L\{f(t)\} - sf(0+) - f'(0+). \tag{1.14}$$

Corollary 1.2. If $f(t), f'(t), \ldots, f^{(n-1)}(t)$ are of exponential order and if $f^{(n-1)}(t)$ is continuous while $f^{(n)}(t)$ is sectionally continuous for $t \geq 0$, then

$$L\{f^{(n)}(t)\} = s^nL\{f(t)\} - s^{n-1}f(0+) - s^{n-2}f'(0+) - \cdots$$
$$- sf^{(n-2)}(0+) - f^{(n-1)}(0+). \tag{1.15}$$

1.5. Inverse Laplace Transforms

If $L\{f(t)\} = F(s)$, then $L^{-1}\{F(s)\} = f(t)$ denotes the function whose transform is $F(s)$. For example,

$$L^{-1}\left\{\frac{1}{s-a}\right\} = e^{at} \qquad (t > 0).$$

$L^{-1}\{F(s)\}$ is called the *inverse Laplace transform*. At this point some comments should be made regarding the uniqueness of inverse transforms. It is known from integration theory that a given function $F(s)$ cannot have more than one inverse transform $f(t)$ that is continuous for all $t \geq 0$. This fact justifies the use of tables. However, not all functions of interest to us will be continuous. For example, if

$f(t) = 1$ for $t > 0$, $F(s) = 1/s$. Thus

$$L^{-1}\left\{\frac{1}{s}\right\} = \begin{cases} 1, & t > 0 \\ 0, & t < 0 \end{cases}$$

is an inverse transform. Because of the discontinuity at $t = 0$, the function

$$f(t) = \begin{cases} 1, & t \geq 0 \\ 0, & t < 0 \end{cases}$$

is another inverse transform. In physical applications the value of the function at the point of discontinuity will be specified by the problem.

In Chapter 4 we are going to study a direct method for finding inverse transforms based on the theory of functions of a complex variable. In the meantime we shall content ourselves with the use of tables, which for most applications is the simplest method anyway. To make use of tables, however, we need a technique for resolving transforms into the forms actually listed. This is the topic of the next section.

Example 1.14. Since

$$L\{e^{bt} \cos at\} = \frac{s - b}{(s - b)^2 + a^2}$$

by the first translation theorem, we have

$$L^{-1}\left\{\frac{s - 2}{(s - 2)^2 + 4}\right\} = e^{2t} \cos 2t.$$

Example 1.15. To find

$$L^{-1}\left\{\frac{5s}{s^2 + 2s + 17}\right\}$$

we proceed as follows:

$$\frac{5s}{s^2 + 2s + 17} = \frac{5s}{(s+1)^2 + 16}$$

$$= 5\frac{s}{(s+1)^2 + 16}$$

$$= 5\frac{(s+1) - 1}{(s+1)^2 + 16}$$

$$= 5\frac{s+1}{(s+1)^2 + 16} - \frac{5}{4}\frac{4}{(s+1)^2 + 16}.$$

Recalling that $L\{\sin at\} = a/(s^2 + a^2)$ and making use of the first translation theorem, we obtain

$$L^{-1}\left\{\frac{5s}{s^2 + 2s + 17}\right\} = 5e^{-t}\cos 4t - \frac{5}{4}e^{-t}\sin 4t.$$

Exercises

1.8. Find

    (a) $L\{te^{2t}\}$

    (b) $L\{3te^{-4t}\}$

1.9. Show that

$$L\{e^{at}t^\alpha\} = \frac{\Gamma(\alpha + 1)}{(s - a)^{\alpha+1}}.$$

1.10. Find

    (a) $L\{e^{bt}\sin at\}$

    (b) $L\{e^{bt}\cosh at\}$

1.11. Given that

$$L\{\sin at\} = \frac{a}{s^2 + a^2},$$

find

(a) $L^{-1}\left\{\dfrac{3}{(s-2)^2+9}\right\}$

(b) $L^{-1}\left\{\dfrac{6}{(s-2)^2+9}\right\}$

1.12. Find

$$L^{-1}\left\{\frac{s}{s^2+6s+13}\right\} \quad \text{and} \quad L^{-1}\left\{\frac{s+1}{s^2+6s+25}\right\}.$$

1.13. Find

$$L^{-1}\left\{\frac{s-1}{s^2-s+3}\right\} \quad \text{and} \quad L^{-1}\left\{\frac{s}{s^2-s+\frac{17}{4}}\right\}.$$

1.14. Using the change of scale theorem, find

$$L^{-1}\left\{\frac{3}{25s^2-4}\right\}.$$

1.15. Given that $L\{\cosh at\} = s/(s^2 - a^2)$, use Theorem 1.4 to find $L\{\sinh at\}$.

1.16. Find an expression for the transform of

$$f''(t) + 2f'(t) - f(t)$$

in terms of $F(s)$ under the additional assumptions that $f(0) = 0$ and $f'(0) = 1$.

1.17. Repeat Exercise 1.16 for

$$2f''(t) - 3f'(t) + f(t), \qquad f(0) = 2, \qquad \text{and} \qquad f'(0) = -3.$$

1.18. Find $L^{-1}\{s/(s^2 + a^2)\}$ by using the infinite-series method.

1.19. Repeat Exercise 1.18 for $L^{-1}\{1/(s - a)\}$.

1.6. Partial Fractions

We know from Section 1.3 that $L^{-1}\{1/s\} = 1$ $(t > 0)$ and $L^{-1}\{1/(s + 1)\} = e^{-t}$ $(t > 0)$. But what about $L^{-1}\{1/[s(s + 1)]\}$?

Now, writing $1/[s(s + 1)]$ in the form $(1/s) - [1/(s + 1)]$, we see that

$$L^{-1}\left\{\frac{1}{s(s + 1)}\right\} = 1 - e^{-t}.$$

More generally, a practical method for finding inverse transforms is obtained by making use of a procedure usually studied in college algebra: A rational function with a numerator of lower degree than the denominator (a *proper* rational function) can be expressed as a sum of *partial fractions* having the following form:

I. To every single factor $as + b$ of the denominator there corresponds a partial fraction $A/(as + b)$, where A is a constant.

II. To every repeating linear factor $(as + b)^n$ of the denominator, there correspond partial fractions of the form

$$\frac{A_1}{as + b} + \frac{A_2}{(as + b)^2} + \cdots + \frac{A_n}{(as + b)^n}.$$

III. To every single quadratic factor $as^2 + bs + c$ of the denominator there corresponds a partial fraction

$$\frac{As + B}{as^2 + bs + c},$$

where A and B are constants.

IV. To every repeating quadratic factor $(as^2 + bs + c)^n$ there correspond partial fractions of the form

$$\frac{A_1s + B_1}{as^2 + bs + c} + \frac{A_2s + B_2}{(as^2 + bs + c)^2} + \cdots + \frac{A_ns + B_n}{(as^2 + bs + c)^n}.$$

Once the forms of the partial fractions are known, the task is reduced to finding the constants appearing in each partial fraction.

Example 1.16. Suppose we reconsider the above transform $1/[s(s + 1)]$. From statement I we see that

$$\frac{1}{s(s + 1)} = \frac{A}{s} + \frac{B}{s + 1},$$

but

$$\frac{A}{s} + \frac{B}{s + 1} = \frac{A(s + 1) + Bs}{s(s + 1)} = \frac{1}{s(s + 1)}.$$

Since the denominators are the same, we must have

$$A(s + 1) + Bs = 1$$

or

$$(A + B)s + A = 1.$$

The last expression is an identity only if the corresponding coefficients are equal, so that $A = 1$ and $A + B = 0$, whence $A = 1$ and $B = -1$.

Example 1.17. As another illustration, consider

$$L^{-1}\left\{ \frac{2s^2 + 7s + 3}{(s + 1)(s - 1)(s + 2)} \right\}.$$

We have

$$\frac{2s^2 + 7s + 3}{(s + 1)(s - 1)(s + 2)}$$

$$= \frac{A}{s + 1} + \frac{B}{s - 1} + \frac{C}{s + 2}$$

$$= \frac{A(s - 1)(s + 2) + B(s + 1)(s + 2) + C(s + 1)(s - 1)}{(s + 1)(s - 1)(s + 2)}.$$

$$(1.16)$$

Consequently,

$$A(s - 1)(s + 2) + B(s + 1)(s + 2) + C(s + 1)(s - 1) = 2s^2 + 7s + 3$$

and

$$(A + B + C)s^2 + (A + 3B)s + (-2A + 2B - C) = 2s^2 + 7s + 3.$$

Again comparing coefficients, we obtain

$$A + B + C = 2,$$
$$A + 3B = 7,$$
$$-2A + 2B - C = 3,$$

from which we get $A = 1$, $B = 2$, and $C = -1$. Hence

$$L^{-1}\left\{\frac{2s^2 + 7s + 3}{(s + 1)(s - 1)(s + 2)}\right\} = L^{-1}\left\{\frac{1}{s + 1} + \frac{2}{s - 1} + \frac{-1}{s + 2}\right\}$$
$$= e^{-t} + 2e^t - e^{-2t}.$$

Although the method of comparing coefficients is direct, it can be laborious. A more efficient alternative, especially for fractions having distinct linear factors in the denominator, is based on the following fact: If two polynomials of degree n are equal for more than n replacements for the variable, they are equal for all values of the variable. Suppose we return to the expression

$$A(s - 1)(s + 2) + B(s + 1)(s + 2) + C(s + 1)(s - 1) = 2s^2 + 7s + 3. \tag{1.17}$$

Because of equation (1.16), equation (1.17) is true for all values of s except possibly $s = 1$, $s = -1$, and $s = -2$, for which the denominators are 0 in equation (1.16). Hence equation (1.17) is true for *all* values of s, *including* 1, -1, and -2. Thus we may set s equal to any one of these values. If $s = 1$,

$$6B = 12 \quad \text{or} \quad B = 2.$$

Similarly, letting $s = -1$ and -2, we get $A = 1$ and $C = -1$, respectively.

Example 1.18. Although tedious in general, comparing coefficients is still the simplest method in some cases. Consider, for example,

$$L^{-1}\left\{\frac{a^2}{s(s^2 + a^2)}\right\}.$$

Since there is a quadratic factor in the denominator, we have, by statement III,

$$\frac{a^2}{s(s^2 + a^2)} = \frac{A}{s} + \frac{Bs + C}{s^2 + a^2} = \frac{A(s^2 + a^2) + (Bs + C)s}{s(s^2 + a^2)}.$$

As before,

$$A(s^2 + a^2) + (Bs + C)s = a^2$$

or

$$(A + B)s^2 + Cs + Aa^2 = a^2.$$

Therefore

$$A + B = 0,$$

$$C = 0,$$

$$Aa^2 = a^2,$$

from which it follows at once that $A = 1$, $C = 0$, and $B = -1$. Thus

$$L^{-1}\left\{\frac{a^2}{s(s^2 + a^2)}\right\} = L^{-1}\left\{\frac{1}{s} - \frac{s}{s^2 + a^2}\right\} = 1 - \cos at.$$

Example 1.19. The function

$$\frac{2a^2s}{s^4 - a^4} = \frac{2a^2s}{(s^2 + a^2)(s - a)(s + a)}$$

also has a quadratic factor in the denominator. Hence

$$\frac{2a^2s}{s^4 - a^4} = \frac{As + B}{s^2 + a^2} + \frac{C}{s - a} + \frac{D}{s + a}$$

and

$$(As + B)(s^2 - a^2) + C(s^2 + a^2)(s + a) + D(s^2 + a^2)(s - a) = 2a^2s.$$

At first glance it appears as though the only convenient values to substitute for s are a and $-a$, but $s^2 + a^2 = 0$ for $s = ai$. So letting $s = ai$ and recalling that two complex numbers are equal if, and only if, their real parts are equal and their imaginary parts are equal, we have

$$(Aai + B)(-2a^2) = 2a^3i$$

or

$$-2Ba^2 - 2Aa^3i = 2a^3i,$$

implying that $A = -1$ and $B = 0$. By setting $s = a$ and $s = -a$, we find that $C = D = \frac{1}{2}$. So

$$L^{-1}\left\{\frac{2a^2s}{s^4 - a^4}\right\} = L^{-1}\left\{\frac{1}{2}\frac{1}{s - a} + \frac{1}{2}\frac{1}{s + a} - \frac{s}{s^2 + a^2}\right\}$$

$$= \tfrac{1}{2}e^{at} + \tfrac{1}{2}e^{-at} - \cos at$$

$$= \cosh at - \cos at.$$

Example 1.20. We have not as yet considered a function with repeating factors in the denominator, as in

$$L^{-1}\left\{\frac{s^2}{(s - 1)^3}\right\}.$$

According to statement II,

$$\frac{s^2}{(s - 1)^3} = \frac{A}{s - 1} + \frac{B}{(s - 1)^2} + \frac{C}{(s - 1)^3}$$

from which

$$A(s - 1)^2 + B(s - 1) + C = s^2. \tag{1.18}$$

Now letting $s = 1$, we obtain $C = 1$, but we seem to have run out of convenient values to substitute. However, remember that whenever two functions are equal, so are their derivatives. Differentiating both sides of equation (1.18), we get

$$2A(s - 1) + B = 2s;$$

now $s = 1$ can be used again, yielding $B = 2$. Differentiating a second time, we obtain

$$2A = 2 \quad \text{or} \quad A = 1.$$

Finally,

$$L^{-1}\left\{\frac{1}{s - 1} + \frac{2}{(s - 1)^2} + \frac{1}{(s - 1)^3}\right\} = e^t + 2te^t + \frac{1}{2}t^2e^t.$$

The differentiation procedure works particularly well with repeating linear factors.

Obviously we could have proceeded by first finding C as before and then substituting two arbitrary values in equation (1.18). For example, letting $s = 0$ and $s = 2$, respectively,

$$A - B + 1 = 0,$$

$$A + B + 1 = 4,$$

whose solution set is $A = 1$ and $B = 2$.

Example 1.21. Find

$$L^{-1}\left\{\frac{3s^3 - 2s^2 + 13s - 6}{(s + 2)(s^2 + 4)^2}\right\}.$$

By statement IV,

$$\frac{3s^3 - 2s^2 + 13s - 6}{(s + 2)(s^2 + 4)^2} = \frac{A}{s + 2} + \frac{Bs + C}{s^2 + 4} + \frac{Ds + E}{(s^2 + 4)^2}$$

and

$$A(s^2 + 4)^2 + (Bs + C)(s + 2)(s^2 + 4) + (Ds + E)(s + 2)$$
$$= 3s^3 - 2s^2 + 13s - 6. \tag{1.19}$$

If $s = -2$,

$$64A = -64 \qquad \text{and} \qquad A = -1.$$

If $s = 2i$,

$$(2Di + E)(2i + 2) = -24i + 8 + 26i - 6$$

or

$$(2E - 4D) + (2E + 4D)i = 2 + 2i.$$

Equating real and imaginary parts, we find that $E = 1$ and $D = 0$.

At this point we could differentiate, as in the last example, but it is no doubt simpler to substitute two arbitrary values into equation (1.19) such as 0 and 1, to obtain $B = C = 1$. Thus

$$L^{-1}\left\{\frac{-1}{s + 2} + \frac{s}{s^2 + 4} + \frac{1}{s^2 + 4} + \frac{1}{(s^2 + 4)^2}\right\}$$

$$= -e^{-2t} + \cos 2t + \frac{1}{2}\sin 2t + \frac{1}{16}\sin 2t - \frac{1}{8}t\cos 2t$$

$$= -e^{-2t} + \cos 2t + \frac{9}{16}\sin 2t - \frac{1}{8}t\cos 2t.$$

The inverse transform of the last term was obtained from the table in Appendix B. A possible way of finding the inverse is through the use of the convolution theorem in Chapter 5. (See Example 5.2 in Chapter 5.)

Exercises

Find the inverse transform in each case.

1.20. $\dfrac{5 - 7s}{(s - 1)(s + 1)(s - 2)}$

1.21. $\dfrac{s^2 + 1}{(s - 1)^3}$

1.22. $\dfrac{s}{(s + 1)^4}$

1.23. $\dfrac{3s^2 + 8s - 1}{(s - 3)(s + 2)^2}$

1.24. $\dfrac{s^2 - 2}{s(s^2 - 4)}$

1.25. $\dfrac{2a^3}{s^4 - a^4}$

1.26. $\dfrac{1}{s^2 - 6s + 13}$

1.27. $\dfrac{a^3}{s^2(s^2 + a^2)}$

1.28. $\dfrac{s^2 - 2s - 2}{(s + 3)(s^2 + 4)}$

1.29. $\dfrac{3s^2 - s + 1}{(s + 1)(s^2 - s + 3)}$

1.30. $\dfrac{s^2 - a^2}{(s^2 + a^2)^2}$

1.31. $\dfrac{3s^4 + 2s^3 + 8s^2 + 7s + 7}{(s - 1)(s^2 + 2)^2}$

1.7. Differential Equations

We saw earlier that linear differential equations with constant coefficients and given initial conditions can be solved by means of Laplace transforms as follows:

(a) Find the Laplace transform of both sides of the differential equation.

(b) Solve the resulting algebraic equation.

(c) Find the inverse transform.

The method is not well suited for solving equations with variable coefficients.

Example 1.22. Solve the differential equation

$$x''(t) + 4x(t) = e^t$$

subject to the initial conditions $x(0) = 0$ and $x'(0) = 1$. We take the transform of both sides, making use of equation (1.14); thus

$$s^2 X(s) - sx(0) - x'(0) + 4X(s) = \frac{1}{s-1}.$$

If we use the initial conditions, this reduces to

$$(s^2 + 4)X(s) = \frac{1}{s-1} + 1$$

or

$$X(s) = \frac{1}{(s^2 + 4)(s - 1)} + \frac{1}{s^2 + 4}.$$

Now split $X(s)$ into partial fractions; that is,

$$X(s) = \frac{1}{5} \frac{1}{s-1} - \frac{1}{5} \frac{s+1}{s^2+4} + \frac{1}{s^2+4}$$

$$= \frac{1}{5} \frac{1}{s-1} - \frac{1}{5} \frac{s}{s^2+4} + \frac{4}{5} \frac{1}{s^2+4}.$$

Finally,

$$x(t) = \frac{1}{5} e^t - \frac{1}{5} \cos 2t + \frac{2}{5} \sin 2t.$$

Example 1.23. Solve the following system of differential equations:

$$x'(t) + 2y'(t) - 2y(t) = t,$$
$$x(t) + y'(t) - y(t) = 1,$$

with initial conditions $x(0) = y(0) = 0$.

The method for solving single differential equations carries over into systems of equations. After transforming each equation and making use of the initial conditions, we obtain

$$sX(s) + 2sY(s) - 2Y(s) = \frac{1}{s^2},$$

$$X(s) + sY(s) - Y(s) = \frac{1}{s},$$

a system of algebraic equations that is readily solved for $X(s)$ and $Y(s)$. From the second equation

$$X(s) = \frac{1}{s} - sY(s) + Y(s).$$

After substituting into the first equation, we have

$$s\left[\frac{1}{s} - sY(s) + Y(s)\right] + 2sY(s) - 2Y(s) = \frac{1}{s^2}$$

or

$$(s^2 - 3s + 2)Y(s) = 1 - \frac{1}{s^2}.$$

Solving, we obtain

$$Y(s) = \frac{s+1}{s^2(s-2)}$$

$$= -\frac{3}{4} \frac{1}{s} - \frac{1}{2} \frac{1}{s^2} + \frac{3}{4} \frac{1}{s-2},$$

and

$$y(t) = -\tfrac{3}{4} - \tfrac{1}{2}t + \tfrac{3}{4}e^{2t}.$$

Normally one would now find $X(s)$ and the resulting inverse transform. In this example, however, it is simpler to substitute $y(t)$ into the second equation to obtain

$$x(t) = \tfrac{3}{4} - \tfrac{3}{4}e^{2t} - \tfrac{1}{2}t.$$

Exercises

Find the solutions of the following differential equations

1.32. $x'(t) + x(t) = 0$, $x(0) = 2$

1.33. $x''(t) + 4x'(t) + 4x(t) = 9e^{t}$, $x(0) = x'(0) = 0$

1.34. $x'(t) + 2x(t) = e^{-2t}$, $x(0) = 1$

1.35. $x''(t) + x(t) = t$, $x(0) = 1$, $x'(0) = 1$

1.36. $x''(t) + x(t) = \sin t$, $x(0) = 1$, $x'(0) = 0$

1.37. $x''(t) + 2x'(t) + 10x(t) = e^{-t} \sin t$, $x(0) = 0$, $x'(0) = 1$

1.38. $x''(t) + 4x'(t) + 4x(t) = 2 \sin t$, $x(0) = 0$, $x'(0) = 1$

1.39. $x'''(t) - 3x''(t) + 3x'(t) - x(t) = te^{t}$,
  $x(0) = 0$, $x'(0) = -1$, $x''(0) = -1$

1.40. $x'(t) - x(t) + 2y(t) = 0$
  $y'(t) + 3x(t) - 2y(t) = 0$
  $x(0) = 3$, $y(0) = 8$

1.41. $3y'(t) - y(t) + 2x'(t) - 3x(t) = 0$
  $2y'(t) + x'(t) - 2x(t) = 2 - 4e^{2t}$
  $x(0) = y(0) = 0$

1.42. $x'(t) + y'(t) - y(t) + x(t) = -2t - 1$
  $x'(t) - y''(t) - 2y(t) = -2t - 5$
  $x(0) = -3$, $y(0) = 3$, $y'(0) = 0$

1.43. $x''(t) + 3x(t) - 2y(t) = 0$
  $x''(t) + y''(t) - 3x(t) + 5y(t) = 0$
  $x(0) = y(0) = 0$, $x'(0) = 1$, $y'(0) = 3$

1.8. Applications

Suppose a mass m is hung on an idealized spring whose upper end is rigidly supported (Figure 1.3). An idealized spring is one for which the mass is negligible and the restoring force is proportional to its extension. Let O, the origin, be the equilibrium position, the x coordinate with downward displacement being positive and upward negative. Suppose that the load is pulled downward a distance x by a force F. Now by Hooke's law $F = -k^2x$, where k^2 is a constant called the force constant of the spring. (We choose k^2 rather than k for convenience of notation in the solution.) Now by Newton's second law

$$F = m\,\frac{d^2x}{dt^2},$$

so that

$$m\,\frac{d^2x}{dt^2} = -k^2x.$$

Finally, assume that a damping force proportional to the velocity is present; that is, the damping force is

$$-b\,\frac{dx}{dt},$$

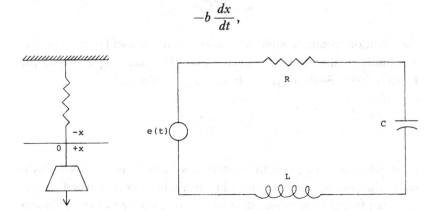

FIGURE 1.3. A mechanical system and its electrical analog.

a fairly accurate assumption for small velocities. Thus

$$m\frac{d^2x}{dt^2} + b\frac{dx}{dt} + k^2x = 0.$$

This is the fundamental equation of *damped simple harmonic motion.*

In discussing the solution of the equation, let $x(0) = X_0$ and $x'(0) = V_0$. Then

$$ms^2X(s) - msX_0 - mV_0 + bsX(s) - bX_0 + k^2X(s) = 0$$

and

$$X(s) = \frac{sX_0 + (b/m)X_0 + V_0}{[s + (b/2m)]^2 + (k^2/m) - (b^2/4m^2)}.$$

Clearly, the nature of the resulting motion depends on the term

$$\frac{k^2}{m} - \frac{b^2}{4m^2}.$$

If the term is positive, or, equivalently, if

$$\frac{k^2}{m} > \frac{b^2}{4m^2},$$

the solution contains sinusoidal terms and is oscillatory, but the amplitude of the oscillation is continuously decreasing owing to the factor $e^{-bt/2m}$. Such a system is called *underdamped.*

If

$$\frac{k^2}{m} < \frac{b^2}{4m^2},$$

no oscillations occur, and the system is said to be *overdamped.* Physically this means that the damping force is large compared to the restoring force of the spring, the mass moving slowly to its equilibrium position without passing it.

Finally, if

$$\frac{k^2}{m} = \frac{b^2}{4m^2},$$

the system is *critically damped*, and, again, the motion is not oscillatory. Physically, the mass comes to its equilibrium position in a minimum time and does not pass beyond its equilibrium position.

As in every oscillatory system, the oscillations cannot be maintained indefinitely due to gradual dissipation of mechanical energy unless energy is supplied to the system. If an external force $f(t)$ is applied, the equation becomes

$$m\frac{d^2x}{dt^2} + b\frac{dx}{dt} + k^2x = f(t).$$

Such cases are called *forced oscillations*.

The system just considered has an electrical analog, as shown in Figure 1.3. By Kirchhoff's voltage law the impressed voltage of a circuit equals the sums of the voltages across the components, so that

$$L\frac{d^2q}{dt^2} + R\frac{dq}{dt} + \frac{q}{C} = e(t).$$

Here q represents the electrical charge and $i = dq/dt$ the current. The three terms on the left give the voltage drop across the inductor, resistor, and capacitor, respectively.

When we compare this equation to the one for the mechanical system it is clear that the inductance L is analogous to the mass m, C to $1/k^2$, and R to the damping force constant b. It should be noted, however, that these analogies are based on the similarity of the equations, not on fundamental concepts, nor are they the only possible set of analogies. Perhaps the most striking feature of the system is that it turns out to be another example of simple harmonic motion.

Exercises

1.44. Suppose that a particle having a mass of 1 gram moves along the x axis and is attracted to the origin by a force equal to $4x$. If the particle is initially at rest at $x = 2$, find its position as a function of time. The equation and initial conditions are

$$x'' + 4x = 0, \qquad x(0) = 2, \qquad \text{and} \qquad x'(0) = 0.$$

1.45. If the particle in Exercise 1.44 experiences a retarding force equal to $4x'$, find its position as a function of time.

1.46. Suppose a falling object experiencing a retarding force due to air resistance has a variable acceleration of $32 - 0.1v$, where v is the velocity. If the object starts from rest, find an expression for the velocity as a function of time. What is the limiting velocity?

1.47. A 64-pound (2-slug) weight stretches a certain spring 2 feet. Suppose the spring is stretched 2 feet beyond its equilibrium position and released. Find the resulting motion.

1.48. Find an expression for the motion in Exercise 1.47 if a damping force numerically equal to the velocity is present.

1.49. Repeat Exercise 1.47 assuming no damping but an external force of $2 \sin 4t$.

1.50. For a given circuit $L = 2$ henrys, $R = 16$ ohms, $C = 0.02$ farad, and $E = 300$ volts. Find the expression relating charge and time if $q(0) = i(0) = 0$.

1.51. Suppose a constant voltage is applied to a general LRC circuit starting at $t = 0$. Find the transform and state conditions under which oscillations occur, that is, when the current becomes alternating.

1.52. Find the well-known formula for the current rise in an inductance,

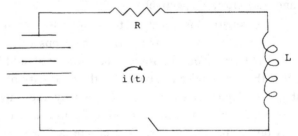

assuming the current to be zero when the switch is closed and E to be a constant (see the diagram).

1.53. Applying Kirchhoff's voltage law to the network in the diagram, we obtain the following system of equations:

$$15i_1 + i_1' - i_2' + 5(i_1 - i_2) = 110,$$
$$10i_2 + 5(i_2 - i_1) + i_2' - i_1' + 2i_2' = 0.$$

Determine the currents given that $i_1(0) = i_2(0) = 0$.

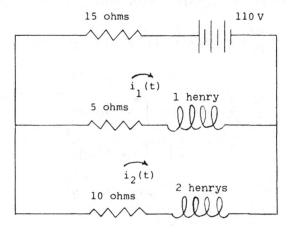

1.54. Consider the transformer circuit in the diagram. Applying Kirchhoff's voltage law, we obtain

$$L_1i_1' + R_1i_1 = Mi_2' + V_0,$$
$$L_2i_2' + R_2i_2 = Mi_1',$$

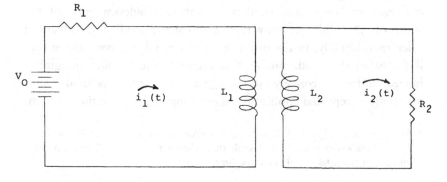

where M is said to be the mutual inductance. If $R_1 = 5$ ohms, $R_2 = 10$ ohms, $L_1 = 1$ henry, $L_2 = 2$ henrys, $M = 1$ henry, and $V_0 = 100$ volts, find $i_2(t)$, the *secondary* current. [Assume that $i_1(0) = i_2(0) = 0$.]

1.9. Differentiation and Integration of Transforms

Theorem 1.6. If $f(t)$ is sectionally continuous and of exponential order $e^{\alpha t}$ and if $L\{f(t)\} = F(s)$, then

$$\frac{d^n}{ds^n} F(s) = L\{(-1)^n t^n f(t)\} \qquad (s > \alpha). \tag{1.20}$$

Proof

$$\frac{d}{ds} F(s) = \frac{d}{ds} \int_0^\infty e^{-st} f(t) \, dt$$

$$= \int_0^\infty \frac{d}{ds} [e^{-st} f(t)] \, dt \tag{1.21}$$

$$= \int_0^\infty e^{-st}[-tf(t)] \, dt$$

$$= L\{-tf(t)\}.$$

The validity of the proof depends on our being able to interchange the order of differentiation and integration in equation (1.21). For improper integrals the procedure is valid provided that the integral *converges uniformly*, that is, that it converges independently of the value of s.* (Uniform convergence is usually studied in advanced calculus.) Actually, in the proof of Theorem 1.3 we have shown that if f satisfies the conditions of this theorem, the Laplace integral is indeed uniformly convergent for $s \geq b > \alpha$. Since this is also true of $L\{t^n f(t)\}$, repeated application of equation (1.21) gives the theorem.

---

* More precisely, $\int_0^\infty g(s, t) \, dt$ converges uniformly on an interval if for every $\varepsilon > 0$ there exists a constant A such that whenever $a > A$, $| \int_a^\infty g(s, t) \, dt | < \varepsilon$ for all s in the interval; A does not depend on s.

Example 1.24

$$L\{te^{at}\} = -\frac{d}{ds} L\{e^{at}\}$$

$$= -\frac{d}{ds} \frac{1}{s-a}$$

$$= \frac{1}{(s-a)^2}.$$

This result also follows from the translation theorem.

We just saw that differentiating the transform of a function corresponds to multiplying the function by $-t$. Similarly, integrating the transform corresponds to dividing the function by t, as shown by the following theorem.

Theorem 1.7. If $f(t)$ is sectionally continuous and of exponential order e^{at} and if $\lim_{t\to 0^+} [f(t)/t]$ exists, then

$$L\left\{\frac{f(t)}{t}\right\} = \int_s^\infty F(x)\,dx. \tag{1.22}$$

Proof. By definition

$$F(s) = \int_0^\infty e^{-st} f(t)\,dt.$$

Integrating both sides, we get

$$\int_s^\infty F(x)\,dx = \int_s^\infty dx \int_0^\infty e^{-xt} f(t)\,dt$$

$$= \lim_{b\to\infty} \int_s^b dx \int_0^\infty e^{-xt} f(t)\,dt$$

$$= \lim_{b\to\infty} \int_0^\infty dt \int_x^b e^{-xt} f(t)\,dx \tag{1.23}$$

$$\int_s^\infty F(x)\,dx = \int_0^\infty \left[\frac{e^{-xt}}{-t}\,f(t)\right]_s^\infty dt = \int_0^\infty e^{-st}\,\frac{f(t)}{t}\,dt \qquad (1.24)$$

$$= L\left\{\frac{f(t)}{t}\right\}.$$

As before, the conditions on $f(t)$ and $f(t)/t$ were introduced to guarantee uniform convergence. Only this way can we be sure that the inter-change of the order of integration in equation (1.23) and the operation of taking the limit under the integral sign in equation (1.24) are actually valid.

Example 1.25

$$L\left\{\frac{\sin at}{t}\right\} = \int_s^\infty \frac{a\,dx}{x^2 + a^2}$$

$$= \text{Arctan}\,\frac{x}{a}\,\Big|_s^\infty$$

$$= \frac{\pi}{2} - \text{Arctan}\,\frac{s}{a}$$

$$= \text{Arccot}\,\frac{s}{a}.$$

An interesting consequence of the theorem is obtained by taking the limit as $s \to 0$ in equation (1.22).

Corollary 1.3. If $F(s) = L\{f(t)\}$, then

$$\int_0^\infty \frac{f(t)}{t}\,dt = \int_0^\infty F(s)\,ds,$$

whenever the integrals exist.

The corollary is useful in evaluating certain improper integrals, as illustrated in the following example.

Example 1.26

$$\int_0^\infty \frac{\sin at}{t}\, dt = \int_0^\infty \frac{a\, ds}{s^2 + a^2}$$

$$= \text{Arctan}\, \frac{s}{a}\Big|_0^\infty$$

$$= \frac{\pi}{2}.$$

The Laplace transform itself may sometimes be useful in evaluating integrals.

Example 1.27

$$\int_0^\infty e^{-t/2} \sin 2t\, dt = \lim_{s \to 1/2} \int_0^\infty e^{-st} \sin 2t\, dt$$

$$= \lim_{s \to 1/2} \frac{2}{s^2 + 4}$$

$$= \frac{8}{17}.$$

Example 1.28. Evaluate

$$\int_0^\infty \frac{e^{-2t} \sinh t}{t}\, dt.$$

$$\int_0^\infty \frac{e^{-st} \sinh t}{t}\, dt = L\left\{\frac{\sinh t}{t}\right\}$$

$$= \int_s^\infty \frac{dx}{x^2 - 1}$$

$$= \frac{1}{2} \ln \frac{x-1}{x+1}\Big|_s^\infty$$

$$= -\frac{1}{2} \ln \frac{s-1}{s+1}.$$

Letting $s = 2$,

$$\int_0^\infty \frac{e^{-2t} \sinh t}{t} \, dt = \frac{1}{2} \ln 3.$$

Exercises

Verify the relationships in Exercises 1.55–1.63.

1.55. (a) $L\{t \sin at\} = \dfrac{2as}{(s^2 + a^2)^2}$

(b) $L\{t \cos at\} = \dfrac{s^2 - a^2}{(s^2 + a^2)^2}$

1.56. $L\{t^2 \cosh at\} = \dfrac{1}{(s - a)^3} + \dfrac{1}{(s + a)^3}$

1.57. (a) $L\left\{\dfrac{\sinh t}{t}\right\} = \dfrac{1}{2} \ln \dfrac{s + 1}{s - 1}$

(b) $L\left\{\dfrac{1 - e^{-t}}{t}\right\} = \ln\left\{1 + \dfrac{1}{s}\right\}$

1.58. $L\left\{\dfrac{\cos nt - \cos mt}{t}\right\} = \dfrac{1}{2} \ln \dfrac{s^2 + m^2}{s^2 + n^2}$

1.59. $\displaystyle\int_0^\infty \dfrac{e^{-bt} \sin at}{t} \, dt = \text{Arctan } \dfrac{a}{b}$

1.60. $\displaystyle\int_0^\infty \dfrac{e^{-t} - e^{-et}}{t} \, dt = 1$

1.61. $\displaystyle\int_0^\infty e^{-t} \cos at \, dt = \dfrac{1}{a^2 + 1}$

1.62. $\displaystyle\int_0^\infty \dfrac{\cos 6t - \cos 12t}{t} \, dt = \ln 2$

1.63. $\displaystyle\int_0^\infty t e^{-3t} \sin t \, dt = 0.06$

1.64. Show that $\displaystyle\int_0^\infty [(\cos at)/t] \, dt$ does not exist.

2

Further Properties and Applications

2.1. The Unit Step Function

Suppose we recall the function in statement (1.11), namely,

$$u(t - t_0) = \begin{cases} 1, & t \geq t_0 \\ 0, & t < t_0, \end{cases} \tag{2.1}$$

to be referred to as the *unit step function* (Figure 2.1). With the aid of the unit step function it is possible to obtain single expressions for functions expressed differently over different intervals and to find

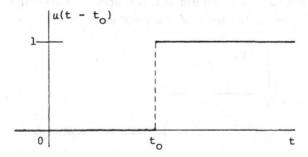

FIGURE 2.1. The unit step function.

43

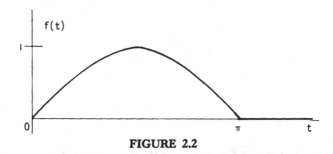

FIGURE 2.2

their Laplace transforms. Suppose, for example, that

$$f(t) = \begin{cases} \sin t, & 0 \le t < \pi \\ 0, & t \ge \pi, \end{cases}$$

whose graph appears in Figure 2.2.

To obtain a single unifying expression for $f(t)$, we consider

$$g(t) = \begin{cases} 1, & 0 \le t < \pi \\ 0, & t \ge \pi, \end{cases}$$

which, in terms of unit step functions, can be written as $g(t) = u(t) - u(t - \pi)$ (Figure 2.3). It follows that

$$f(t) = [u(t) - u(t - \pi)] \sin t. \tag{2.2}$$

To find the Laplace transform of this function we need an additional theorem. This is the topic of the next section.

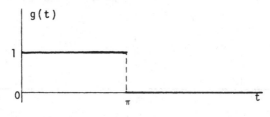

FIGURE 2.3

2.2. The Second Translation Theorem

Theorem 2.1. Second Translation Theorem. If $F(s) = L\{f(t)\}$ exists, then

$$L\{u(t - c)f(t - c)\} = e^{-cs}F(s) \qquad (c \geq 0). \tag{2.3}$$

Proof. The proof follows immediately from the definition of the unit step function; thus

$$L\{u(t - c)f(t - c)\} = \int_0^\infty e^{-st}u(t - c)f(t - c)\, dt$$

$$= \int_c^\infty e^{-st} f(t - c)\, dt.$$

Making the change of variable $u = t - c$, we get

$$\int_0^\infty e^{-s(u+c)} f(u)\, du = e^{-cs} \int_0^\infty e^{-su} f(u)\, du$$

$$= e^{-cs}F(s).$$

Example 2.1. Suppose we apply the second translation theorem to the unit step function:

$$L\{u(t - t_0)\} = L\{u(t - t_0) \cdot 1\}$$

$$= e^{-t_0 s} \cdot \frac{1}{s}$$

$$= \frac{e^{-t_0 s}}{s},$$

or

$$L\{u(t - t_0)\} = \frac{e^{-st_0}}{s} \qquad (s > 0). \tag{2.4}$$

In particular, for $t_0 = 0$,

$$L\{u(t)\} = \frac{1}{s} \qquad (s > 0). \tag{2.5}$$

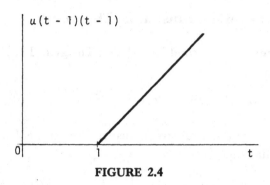

FIGURE 2.4

Example 2.2. By the theorem,

$$L\{u(t-1)(t-1)\} = \frac{e^{-s}}{s^2}.$$

(See Figure 2.4.)

Example 2.3. Suppose we wish to find the transform of $f(t) = [u(t) - u(t-\pi)]\sin t$, considered in Section 2.1. To employ the theorem we must first express $\sin t$ as a function of $t - \pi$. From trigonometry,

$$\sin(t - \pi) = \sin t \cos \pi - \cos t \sin \pi = -\sin t.$$

Hence

$$L\{[u(t) - u(t-\pi)]\sin t\} = L\{u(t)\sin t + u(t-\pi)\sin(t-\pi)\}$$

$$= \frac{1}{s^2 + 1} + \frac{e^{-\pi s}}{s^2 + 1}$$

$$= \frac{1 + e^{-\pi s}}{s^2 + 1}.$$

Example 2.4. Find $L\{u(t-2)f(t)\}$, where

$$f(t) = t^2 - 2t + 4.$$

To express $f(t)$ as a function of $t - 2$, we let

$$t^2 - 2t + 4 = A(t - 2)^2 + B(t - 2) + C$$

and compare coefficients by writing the right-hand side in the form

$$t^2 - 2t + 4 = At^2 + (B - 4A)t + (4A - 2B + C).$$

Now $A = 1$, and since $B - 4A = -2$, we have $B = 2$; from $4A - 2B + C = 4$ it follows that $C = 4$. Consequently,

$$f(t) = (t - 2)^2 + 2(t - 2) + 4$$

and

$$L\{u(t - 2)f(t)\} = L\{u(t-2)(t-2)^2 + 2u(t-2)(t-2) + 4u(t-2)\}$$

$$= \frac{2e^{-2s}}{s^3} + \frac{2e^{-2s}}{s^2} + \frac{4e^{-2s}}{s}$$

$$= \frac{e^{-2s}}{s}\left(\frac{2}{s^2} + \frac{2}{s} + 4\right).$$

For more complicated polynomials the following observation may be helpful: Any polynomial can be viewed as a function written in the form of a Maclaurin series. To write the polynomial as a function of $t - a$, expand it in a Taylor series about a. For example, if $f(t) = 5t^5 + 10t^4 - 2t^3 + t^2 + 5$ is to be expressed as a function of $t - 1$, find the five nonzero derivatives at $t = 1$ and use Taylor's formula. A simple calculation gives

$$f(t) = 19 + 61(t - 1) + \frac{210}{2!}(t - 1)^2 + \frac{528}{3!}(t - 1)^3$$

$$+ \frac{840}{4!}(t - 1)^4 + \frac{600}{5!}(t - 1)^5.$$

When finding the transform of $u(t - 1)f(t)$, the factorials cancel,

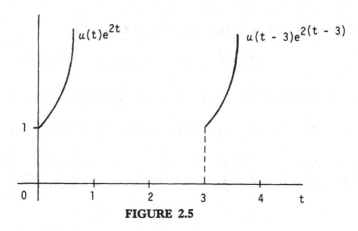

FIGURE 2.5

and one simply has

$$e^{-s}\left(\frac{19}{s} + \frac{61}{s^2} + \frac{210}{s^3} + \frac{528}{s^4} + \frac{840}{s^5} + \frac{600}{s^6}\right).$$

Example 2.5. Find the inverse transform of

$$F(s) = \frac{e^{-3s}}{s-2}.$$

By the theorem we obtain directly

$$f(t) = u(t - 3)e^{2(t-3)}.$$

In sketching the graph the student should note that the graph of $e^{2(t-3)}$ may be obtained from the graph of e^{2t} by translating the latter three units to the right (Figure 2.5).

Exercises

In Exercises 2.1–2.4, write in terms of unit functions and sketch.

2.1. $f(t) = \begin{cases} 1, & 0 \le t < \pi \\ 0, & t \ge \pi \end{cases}$

2.2. $f(t) = \begin{cases} 0, & 0 \leq t < 1 \\ 2, & 1 \leq t < 3 \\ 0, & t \geq 3 \end{cases}$

2.3. $f(t) = \begin{cases} 0, & 0 \leq t < 2\pi \\ \cos t, & t \geq 2\pi \end{cases}$

2.4. $f(t) = \begin{cases} 0, & 0 \leq t < 2 \\ t^2, & 2 \leq t < 3 \\ 0, & t \geq 3 \end{cases}$

2.5. Find $L\{u(t-3)(t-3)^3\}$.

2.6. Find

    (a) $\quad L\{u(t-\pi)\cos(t-\pi)\}$

    (b) $\quad L\{u(t-2\pi)\cos 2t\}$

2.7. Find $L\{u(t-3)(2t^2 + 3t - 1)\}$.

2.8. Use Taylor's formula to find

$$L\{u(t-2)(3t^4 - 2t^3 + 5t^2 + t - 1)\}.$$

2.9. If $f(t) = \sum_{k=0}^{n} a_k t^k$, a polynomial of degree n, show that

$$L\{u(t-c)f(t)\} = e^{-cs} \sum_{k=0}^{n} \frac{f^{(k)}(c)}{s^{k+1}}$$

[as usual, $f^{(0)}(t) = f(t)$].

In Exercises 2.10–2.14, find the inverse transform and sketch.

2.10. $L^{-1}\left\{\dfrac{e^{-5s}}{s^2}\right\}$

2.11. $L^{-1}\left\{\dfrac{1 - e^{-3s}}{s}\right\}$

2.12. $L^{-1}\left\{\dfrac{(1 - e^{-s})^2}{s^3}\right\}$

2.13. $L^{-1}\left\{\dfrac{se^{-\pi s}}{s^2 + 4}\right\}$

2.14. $L^{-1}\left\{\dfrac{e^{-as}}{s^2 - 1}\right\}$

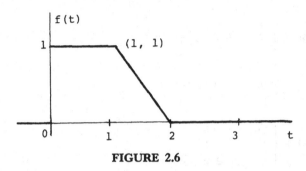

FIGURE 2.6

2.3. Transforms by Graphical Addition

Transforms of certain types of functions are often obtained by graphical addition, as illustrated by the following examples.

Example 2.6. Find $L\{f(t)\}$, where $f(t)$ is the function whose graph appears in Figure 2.6.

We recall from the last section that the horizontal segment can be written $u(t) - u(t - 1)$. Now the segment joining $(1, 1)$ and $(2, 0)$ is part of the graph of $-t$ translated 2 units to the right to become $-(t - 2)$. Consequently,

$$u(t) - u(t - 1) + u(t - 1)[-(t - 2)]$$

is the function whose graph appears in Figure 2.7. Finally, adding

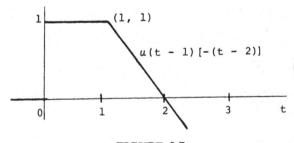

FIGURE 2.7

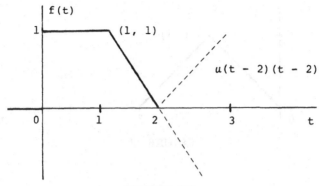

FIGURE 2.8

$u(t-2)(t-2)$ (Figure 2.8), we get

$$f(t) = u(t) - u(t-1) + u(t-1)[-(t-2)] + u(t-2)(t-2).$$

To apply Theorem 2.1 we need to convert the form of $f(t)$. In particular,

$$u(t-1)[-(t-2)] = u(t-1)[-(t-1-1)]$$
$$= -u(t-1)(t-1) + u(t-1).$$

Hence

$$f(t) = u(t) - u(t-1)(t-1) + u(t-2)(t-2)$$

and

$$L\{f(t)\} = \frac{1}{s} - \frac{e^{-s}}{s^2} + \frac{e^{-2s}}{s^2}.$$

Example 2.7. Find $L\{f(t)\}$ for $f(t)$ in Figure 2.9. The ray through $(0, 0)$ and $(1, 1)$ can be written as $u(t)t$. Subtracting $2u(t-1)$ $\times (t-1)$, we obtain the function whose graph is shown in Figure 2.10. Finally, adding $u(t-2)(t-2)$, we have

$$f(t) = u(t)t - 2u(t-1)(t-1) + u(t-2)(t-2). \qquad (2.6)$$

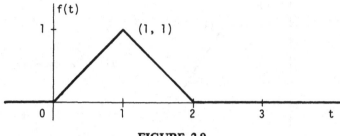

FIGURE 2.9

Hence

$$L\{f(t)\} = \frac{1}{s^2} - \frac{2e^{-s}}{s^2} + \frac{e^{-2s}}{s^2}$$

$$= \frac{1}{s^2}\,(1 - 2e^{-s} + e^{-2s})$$

$$= \frac{1}{s^2}\,(1 - e^{-s})^2.$$

An alternative procedure is to write the segment joining $(0, 0)$ and $(1, 1)$ in the form

$$[u(t) - u(t - 1)]t;$$

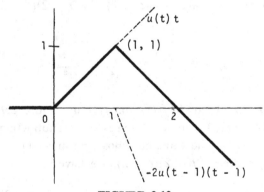

FIGURE 2.10

similarly,

$$[u(t-1) - u(t-2)][-(t-2)]$$

is the segment joining $(1, 1)$ and $(2, 0)$. Adding,

$$f(t) = [u(t) - u(t-1)]t + [u(t-1) - u(t-2)][-(t-2)],$$

which reduces to equation (2.6).

Exercises

2.15. Find the transforms of the functions in the diagrams.

(a)

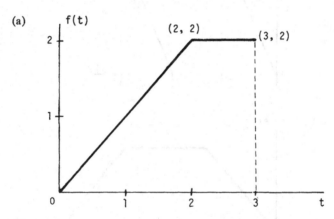

(b)

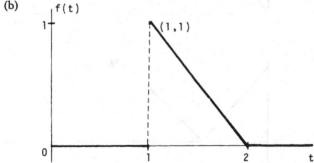

(c)

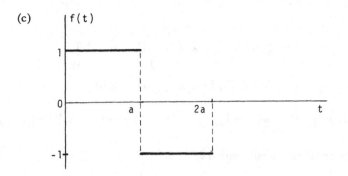

(d)

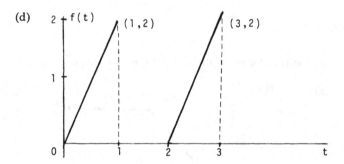

(e)

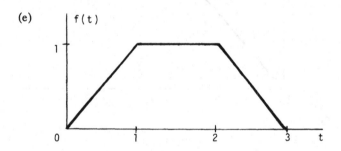

(f)

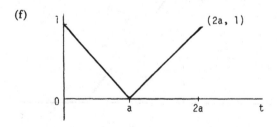

2.4. The Unit Impulse Function

Suppose we consider the function $f(t)$ shown in Figure 2.11. Since the function was constructed so that the area under the graph is 1,

$$\int_{-\infty}^{\infty} f(t)\, dt = 1. \tag{2.7}$$

If we now let ε go to 0, the function becomes 0 everywhere except at t_0, where it is undefined. Denoting the limiting function by $\delta(t - t_0)$ and using equation (2.7), we get

$$\lim_{\varepsilon \to 0} \int_{-\infty}^{\infty} f(t)\, dt = \int_{-\infty}^{\infty} \lim_{\varepsilon \to 0} f(t)\, dt \tag{2.8}$$

$$= \int_{-\infty}^{\infty} \delta(t - t_0)\, dt$$

$$= 1.$$

In other words, $\delta(t - t_0)$, called the *Dirac delta function* or *unit impulse function*, is defined as

$$\delta(t - t_0) = \begin{cases} 0, & t \neq t_0 \\ \infty, & t = t_0 \end{cases}$$

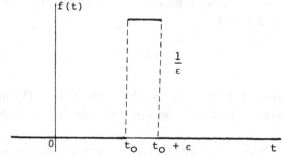

FIGURE 2.11. A pre-limit form of the unit impulse function.

and such that

$$\int_{-\infty}^{\infty} \delta(t - t_0)\, dt = 1.$$

Now, mathematically speaking, this definition is absurd. The integral of any function that is 0 except at a single point has a value of 0. The problem lies in step (2.8), where we interchanged two operations, that of performing the integration and that of taking the limit. We ran into the same difficulty in Section 1.9 on the integration of transforms, the difference being that in the earlier section the step was valid, whereas here it is not. Yet operations involving delta functions have proved to be so useful that the "function" has survived in spite of the mathematical objections. We note in passing that these objections were finally overcome in 1951 by the French mathematician L. Schwartz,[*] who showed that our $\delta(t - t_0)$ is a special case of a *distribution* or *generalized function*, a highly useful concept in both pure and applied mathematics. Distributions, as it turns out, are viewed as functions not on the real line, but on certain sets. Thus generalized functions became a fruitful extension of the classical function concept.

From a physical standpoint $\delta(t - t_0)$ is interpreted as a blow, such as the blow of a hammer acting only at the instant t_0 or as a concentrated load acting at a single point t_0. To make this interpretation plausible, suppose a beam extends between points $-a$ and a along the x axis. If $F(x)$ represents the load density between $-a$ and a, the total load is

$$\int_{-a}^{a} F(x)\, dx.$$

Suppose a body having a trapezoidal cross section (Figure 2.12) extends between $-b$ and b. The load intensity is zero from $-a$ to

---

[*] L. Schwartz, "Théorie des distributions," *Actualités scientifique et industrielles*, Nos. 1091 and 1122, Hermann & Cie, Paris, 1950–1951.

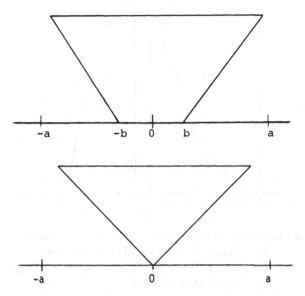

FIGURE 2.12. Physical interpretation of the impulse function as a concentrated load.

$-b$ and from b to a but different from zero between $-b$ and b. The total load is still the integral of the load density between $-a$ and a. If the weight of the body is kept the same but the base is shrunk, the load density increases, compensating for the reduction of the loaded interval. In the limit the load will be concentrated at 0, where the load density becomes infinitely large. At all other points the density is zero, and the integral, which is the weight of the body, remains constant.

If a particle is given a blow, the force varies greatly and lasts for only a short time. It starts at zero, rises sharply to a maximum (which may be quite large), and rapidly falls to zero again (Figure 2.13). From Newton's second law

$$f \, dt = d(mv),$$

so that $f \, dt$ is the elementary change in momentum. The integral

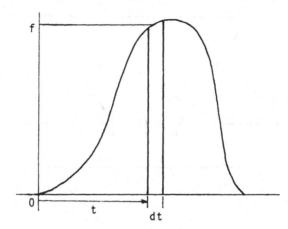

FIGURE 2.13. Physical interpretation of the impulse function as a large force acting on a vanishingly small interval; the area under the graph is the change in momentum of the particle, also called the "strength" of the impulse.

$\int_0^t f \, dt$, the area under the curve, is the total change in momentum of the particle, for if v_0 is the velocity at the time $t = 0$ when the force starts to act and v the velocity at time t when the force ceases to act,

$$\int_0^t f \, dt = \int_{v_0}^v d(mv) = mv - mv_0.$$

The integral $\int_0^t f \, dt$ is called the *impulse* of the force. Even though we are dealing with a very complicated event, the form of the impulse curve need not concern us. (It depends on the nature of the impact.) As before, if the impulse has a certain magnitude, say unity, and the interval $[0, t]$ during which the blow lasts shrinks to zero, f becomes the δ function. The meaning of the δ function in this situation now becomes clear: It represents a large force acting on a vanishingly small time interval, while

$$\int_{-\infty}^\infty \delta(t) \, dt = 1$$

is the *strength* of the impulse, that is, the change in momentum of the particle. These examples explain why *one never talks about the "value" of $\delta(t)$ but only of the value of the integral involving $\delta(t)$.* It is also worth noting that the δ function can be used in problems on electrical circuits.

To obtain the Laplace transform of $\delta(t - t_0)$, we return to Figure 2.11, write the function in terms of unit step functions, and let $\varepsilon \to 0$. Clearly,

$$f(t) = \frac{1}{\varepsilon} \{ u(t - t_0) - u[t - (t_0 + \varepsilon)] \}$$

and

$$L\{f(t)\} = \frac{1}{\varepsilon} \left(\frac{e^{-st_0}}{s} - \frac{e^{-(t_0 + \varepsilon)s}}{s} \right)$$

$$= \frac{e^{-st_0}(1 - e^{-s\varepsilon})}{s\varepsilon}.$$

Now by L'Hospital's rule,

$$\lim_{\varepsilon \to 0} \frac{e^{-st_0}(1 - e^{-s\varepsilon})}{s\varepsilon} = e^{-st_0},$$

so that

$$L\{\delta(t - t_0)\} = e^{-st_0}.$$

If $t_0 = 0$, we get the striking formula

$$L\{\delta(t)\} = 1.$$

By a similar procedure one can show that, more generally, for a continuous function $g(t)$

$$\int_{-\infty}^{\infty} g(t)\delta(t - t_0) \, dt = g(t_0). \tag{2.9}$$

Derivatives of generalized functions were also defined by Schwartz, justifying the formula

$$\frac{d}{dt}\,u(t) = \delta(t). \qquad (2.10)$$

(See Example 2.10 in the next section.) This formula is, in one sense, intuitively obvious: The slope of $u(t)$ is zero except at $t = 0$, where it is infinitely large.

As an example, consider the differential equation

$$x'(t) + x(t) = \delta(t - a), \qquad x(0) = 0.$$

Transforming both sides of the equation,

$$sX(s) + X(s) = e^{-as}$$

and

$$X(s) = \frac{e^{-as}}{s + 1}.$$

Hence

$$x(t) = u(t - a)e^{-(t-a)}.$$

As another example, suppose the mass m on a spring is initially at rest but struck from above by a blow of strength P at $t = 0$. Assuming no damping,

$$mx''(t) + k^2x(t) = P\delta(t), \qquad x(0) = x'(0) = 0.$$

From the transformed equation

$$ms^2X(s) + k^2X(s) = P,$$

$$X(s) = \frac{P}{ms^2 + k^2}$$

and

$$x(t) = \frac{P}{km^{1/2}} \sin \frac{k}{m^{1/2}} t.$$

To see that P is indeed the change in momentum of the mass, note that

$$mx'(t) = P \cos \frac{k}{m^{1/2}} t \to P \qquad \text{as } t \to 0^+.$$

So if the mass was initially at rest, its momentum has jumped to P at the instant $t = 0$. Only in this sense, incidently, are the initial conditions satisfied.

The equation itself is obviously satisfied by the solution obtained if $t > 0$. A direct check using the usual form

$$x(t) = \frac{P}{km^{1/2}} u(t) \sin \frac{k}{m^{1/2}} t$$

is somewhat more difficult. However,

$$x'(t) = \frac{P}{m} u(t) \cos \frac{k}{m^{1/2}} t + \frac{P}{km^{1/2}} \delta(t) \sin \frac{k}{m^{1/2}} t$$

by formula (2.10) and the product rule. Now, according to equation (2.9), for any continuous function $g(t)$

$$\int_{-\infty}^{\infty} \delta(t) g(t) \sin \frac{k}{m^{1/2}} t \, dt = g(0) \sin 0 = 0,$$

which is possible only if $\delta(t) \sin(k/m^{1/2})t = 0$. Hence

$$x'(t) = \frac{P}{m} u(t) \cos \frac{k}{m^{1/2}} t.$$

Since

$$x''(t) = -\frac{Pk}{mm^{1/2}}\, u(t) \sin \frac{k}{m^{1/2}}\, t + \frac{P}{m}\, \delta(t) \cos \frac{k}{m^{1/2}}\, t$$

$$= -\frac{Pk}{mm^{1/2}}\, u(t) \sin \frac{k}{m^{1/2}}\, t + \frac{P}{m}\, \delta(t),$$

it now follows directly that

$$mx''(t) + k^2 x(t) = P\delta(t).$$

For the corresponding electrical analog,

$$L\frac{d^2q(t)}{dt^2} + \frac{1}{C}\, q(t) = E_0\delta(t), \qquad q(0) = i(0) = 0,$$

see Exercise 2.32. Here $E_0\delta(t)$ is viewed as a voltage impulse of strength E_0 applied to the circuit at $t = 0$. It is easy to check that at this instant the current jumps from 0 to E_0/L.

Another instructive exercise is to solve the above spring problem by using the function in Figure 2.11 for $t_0 = 0$ and then letting $\varepsilon \to 0$. Thus

$$mx''(t) + k^2 x(t) = \frac{P}{\varepsilon}\, [u(t) - u(t - \varepsilon)], \qquad x(0) = x'(0) = 0.$$

Transforming, we obtain

$$ms^2 X(s) + k^2 X(s) = \frac{P}{\varepsilon}\left(\frac{1}{s} - \frac{e^{-s\varepsilon}}{s}\right),$$

whence

$$X(s) = \frac{P}{m\varepsilon}\, \frac{1}{s(s^2 + k^2/m)}\, (1 - e^{-s\varepsilon})$$

$$= \frac{P}{k^2\varepsilon}\left(\frac{1}{s} - \frac{s}{s^2 + k^2/m}\right)(1 - e^{-s\varepsilon})$$

and

$$x(t) = \frac{P}{k^2 \varepsilon} \left\{ u(t) \left(1 - \cos \frac{k}{m^{1/2}} t \right) - u(t - \varepsilon) \left[1 - \cos \frac{k}{m^{1/2}} (t - \varepsilon) \right] \right\}.$$

If we now let $\varepsilon \to 0$ by using L'Hospital's rule again,

$$x(t) = \frac{P}{km^{1/2}} \sin \frac{k}{m^{1/2}} t,$$

the solution obtained earlier. Finally, as long as $\varepsilon > 0$, that is, as long as the impulse is assumed to have a small but nonzero duration, the initial conditions are satisfied.

2.5. Applications

Example 2.8. An inductor of L henrys and a capacitor of C farads are connected in series with a generator of $e(t)$ volts. Find the charge q as a function of time if $q(0) = 0$, $i(0) = 0$, and

$$e(t) = \begin{cases} 0, & 0 \le t < a \\ E_0, & t \ge a. \end{cases}$$

The equation is

$$L \frac{d^2 q(t)}{dt^2} + \frac{1}{C} q(t) = E_0 u(t - a).$$

Taking Laplace transforms,

$$Ls^2 Q(s) + \frac{1}{C} Q(s) = \frac{E_0 e^{-as}}{s},$$

and

$$Q(s) = \frac{E_0}{L} \frac{e^{-as}}{s(s^2 + 1/LC)}$$

$$= \frac{E_0 C e^{-as}}{s} - \frac{s E_0 C e^{-as}}{s^2 + 1/LC}.$$

Hence

$$q(t) = E_0 C u(t - a) \left[1 - \cos \frac{t - a}{(LC)^{1/2}} \right]$$

or

$$q(t) = \begin{cases} 0, & 0 \le t < a \\ E_0 C \left[1 - \cos \dfrac{t - a}{(LC)^{1/2}} \right], & t \ge a; \end{cases}$$

$q(t)$ is continuous with a continuous first derivative.*

Remark 2.1. This example illustrates the real power of the Laplace transform. The fact that $e(t)$ is not continuous poses a considerable problem for the older classical methods, as would a function not possessing a continuous derivative (such as the function shown in Figure 2.2). Using Laplace transforms, either case can be handled with ease.

Another field to which the unit function and impulse function may be applied is the study of deflection of beams, where discontinuous and concentrated loads are encountered.

Suppose the beam is placed along the x axis with the left end at the origin. Since the vertical load $F(x)$ per unit length is assumed to act on the beam in the downward direction, it is natural to consider the downward deflection $Y(x)$ from this line positive (Figure 2.14).

The deflection is a function of the length x and is known to satisfy the differential equation

$$\frac{d^4 Y}{dx^4} = \frac{F(x)}{EI}, \tag{2.11}$$

---

* To check the solution, note that $q''(a+)$ exists and is equal to

$$\lim_{t \to a+} q''(t) = \frac{E_0}{L} \cos \frac{t - a}{(LC)^{1/2}} \bigg|_{t=a} = \frac{E_0}{L}$$

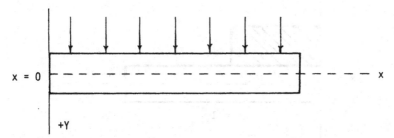

FIGURE 2.14. A beam with a load acting in the downward direction.

where $-EI(d^2Y/dx^2)$ is the internal bending moment and $-EI(d^3Y/dx^3)$ the shearing force. E is Young's *modulus of elasticity* and I the moment of inertia of a cross section about the axis. The product EI is called the *flexural rigidity* of the beam.

In solving the equation by the transform method, the transform is obviously taken with respect to the space variable x.* For the boundary conditions we shall confine ourselves to two cases:

$$
\begin{array}{ll}
\text{1.} & \text{Clamped end: } Y = Y' = 0, \\
\text{2.} & \text{Free end: } Y'' = Y''' = 0.
\end{array}
\tag{2.12}
$$

Example 2.9. Suppose the beam in Figure 2.15 is 2 units long, clamped at $x = 0$ and free at $x = 2$. Assume that the load is given by

$$F(x) = F_0[\omega(x) - \omega(x - 1)].$$

From the discussion above,

$$\frac{d^4Y}{dx^4} = \frac{F_0}{EI}[\omega(x) - \omega(x - 1)]$$

and $Y(0) = 0$, $Y'(0) = 0$, $Y''(2) = 0$, and $Y'''(2) = 0$. Then by

* We shall use the notation $y(s) = L\{Y(x)\}$.

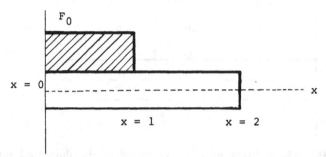

FIGURE 2.15. A cantilever beam, clamped at the left end and free at the right end, carrying a uniform load F_0 per unit length on its left half.

equation (1.15),

$$s^4 y(s) - s^3 Y(0) - s^2 Y'(0) - s Y''(0) - Y'''(0) = \frac{F_0}{EI} \frac{1 - e^{-s}}{s}.$$

Now $Y(0) = Y'(0) = 0$, but $Y''(0)$ and $Y'''(0)$ are not known and will have to be carried along as arbitrary constants to be determined later. Solving for $y(s)$,

$$y(s) = \frac{Y''(0)}{s^3} + \frac{Y'''(0)}{s^4} + \frac{F_0}{EIs^5} (1 - e^{-s}),$$

so that

$$Y(x) = \frac{Y''(0)x^2}{2!} + \frac{Y'''(0)x^3}{3!} + \frac{F_0 x^4}{4! \, EI} - \frac{F_0}{EI} u(x-1) \frac{(x-1)^4}{4!}.$$

On the interval $(1, 2)$

$$Y(x) = \frac{Y''(0)x^2}{2!} + \frac{Y'''(0)x^3}{3!} + \frac{F_0 x^4}{4! \, EI} - \frac{F_0}{EI} \frac{(x-1)^4}{4!}.$$

To use the conditions $Y''(2) = Y'''(2) = 0$, we find

$$Y''(x) = Y''(0) + Y'''(0)x + \frac{F_0 x^2}{2EI} - \frac{F_0(x-1)^2}{2EI}$$

and

$$Y'''(x) = Y'''(0) + \frac{F_0 x}{EI} - \frac{F_0(x-1)}{EI}, \qquad 1 \le x < 2,$$

and let $x = 2$. Hence

$$Y''(0) = \frac{F_0}{2EI} \qquad \text{and} \qquad Y'''(0) = -\frac{F_0}{EI}.$$

Consequently, the deflection is

$$Y(x) = \frac{F_0 x^2}{4EI} - \frac{F_0 x^3}{6EI} + \frac{F_0 x^4}{24EI} - \frac{F_0}{24EI}\, u(x-1)(x-1)^4.$$

Example 2.10. A particle of mass m is initially at rest and is set in motion by a blow of strength P at $t = 0$. Thus

$$m\frac{d^2x}{dt^2} = P\delta(t);$$

transforming,

$$ms^2 X(s) = P,$$

$$X(s) = \frac{P}{ms^2},$$

and

$$x(t) = u(t)\frac{Pt}{m}.$$

Equivalently,

$$m\frac{dv}{dt} = P\delta(t), \qquad (2.13)$$

$$msV(s) = P,$$

$$V(s) = \frac{P}{ms},$$

and

$$v(t) = \frac{P}{m}\, u(t)$$

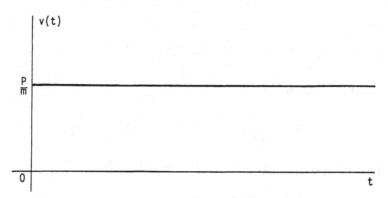

FIGURE 2.16. Response of a particle of mass m to a blow of strength P.

(See Figure 2.16.) Substituting the second solution into equation (2.13), we obtain $(d/dt)\varkappa(t) = \delta(t)$, confirming formula (2.10).

Example 2.11. A beam of length a is clamped at both ends and carries a concentrated load P_0 at its midpoint (see Figure 2.17). Find the deflection.

The concentrated load can be represented by the impulse function $P_0\delta(t - \tfrac{1}{2}a)$. From equations (2.11) and (2.12),

$$\frac{d^4Y}{dx^4} = \frac{P_0}{EI}\,\delta(x - \tfrac{1}{2}a),$$

$Y(0) = Y'(0) = 0$ and $Y'(a) = Y(a) = 0$.

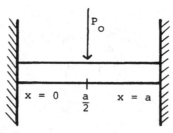

FIGURE 2.17. A beam clamped at both ends with a concentrated load P_0 acting downward at the midpoint.

Taking Laplace transforms, we have

$$s^4 y(s) - s^3 Y(0) - s^2 Y'(0) - s Y''(0) - Y'''(0) = \frac{P_0}{EI} e^{-as/2}.$$

As before, $Y''(0)$ and $Y'''(0)$ must be determined later. Using the other two conditions, we obtain

$$y(s) = \frac{Y''(0)}{s^3} + \frac{Y'''(0)}{s^4} + \frac{P_0}{EI} \frac{e^{-as/2}}{s^4}$$

and

$$Y(x) = \frac{Y''(0)x^2}{2!} + \frac{Y'''(0)x^3}{3!} + \frac{P_0}{EI} u\left(x - \frac{a}{2}\right) \frac{(x - \frac{1}{2}a)^3}{3!}.$$

On the interval $(\frac{1}{2}a, a)$,

$$Y(x) = \frac{Y''(0)x^2}{2!} + \frac{Y'''(0)x^3}{3!} + \frac{P_0}{EI} \frac{(x - \frac{1}{2}a)^3}{3!}$$

and

$$Y'(x) = Y''(0)x + \frac{Y'''(0)x^2}{2} + \frac{P_0}{EI} \frac{(x - \frac{1}{2}a)^2}{2}.$$

Letting $x = a$, we find that

$$Y''(0) = \frac{P_0 a}{8EI} \quad \text{and} \quad Y'''(0) = -\frac{P_0}{2EI},$$

so that the solution is given by

$$Y(x) = \frac{P_0}{EI}\left[\frac{ax^2}{16} - \frac{x^3}{12} + \frac{1}{6} u\left(x - \frac{1}{2}a\right)\left(x - \frac{1}{2}a\right)^3\right].$$

In Exercise 2.40 we are going to need the

Hurwitz Criterion. Given the real polynomial

$$F(x) = x^n + a_1 x^{n-1} + \cdots + a_n,$$

form the determinants $A_1 = a_1$ and

$$
A_k = \begin{vmatrix}
a_1 & a_3 & a_5 & \cdots & a_{2k-1} \\
1 & a_2 & a_4 & \cdots & a_{2k-2} \\
0 & a_1 & a_3 & \cdots & a_{2k-3} \\
0 & 1 & a_2 & \cdots & a_{2k-4} \\
& & \cdots & & \\
0 & 0 & 0 & \cdots & a_k
\end{vmatrix}
$$

for $k = 2, 3, \ldots, n$ with $a_j = 0$ for $j > n$. If all the determinants A_k are positive, $F(x)$ has only zeros with negative real parts.

Exercises

Exercises 2.16–2.28 involve unit step functions.

2.16. The weight on a spring is pulled b units beyond its equilibrium position and released. Find the resulting motion if no damping is present but a force f is acting on the spring, where

$$
f = \begin{cases} 0, & 0 \le t < a \\ F_0, & t \ge a. \end{cases}
$$

The equation is

$$
mx'' + k^2 x = f, \qquad x(0) = b, \quad x'(0) = 0.
$$

2.17. In Example 2.8 let $L = 2$ henrys, $C = 0.01$ farad, and

$$
e(t) = \begin{cases} 100t, & 0 \le t < 1 \\ 0, & t \ge 1. \end{cases}
$$

Find $q(t)$ if $q(0) = i(0) = 0$.

2.18. An EMF [electromotive force (volts)] is applied to a resistance and inductance in series. Find an expression for the current as a function of time, assuming that the current is 0 when the EMF is first applied, if

(a) $e(t) = u(t)E_0$

(b) $e(t) = \begin{cases} E_0, & 0 \le t < a \\ 0, & t \ge a \end{cases}$

2.19. The weight on a spring is initially at rest. Suppose a force f, where

$$f = \begin{cases} 1, & 0 \le t < \pi/k \\ 0, & t \ge \pi/k, \end{cases}$$

is acting on the spring but no damping is present. For $m = 1$, the equation and boundary conditions become

$$x'' + k^2 x = f, \qquad x(0) = x'(0) = 0.$$

Find the resulting motion.

2.20. Suppose the weight in Exercise 2.19 is at the equilibrium position when $t = 0$ but has an initial velocity X_1. Find the motion if

$$f = \begin{cases} 1, & 0 \le t < \pi/k \\ 0, & t \ge \pi/k. \end{cases}$$

2.21. The weight on the spring in Exercise 2.19 is initially at rest, and the force acting on the spring is given by

$$f(t) = \begin{cases} \sin t, & 0 \le t < \pi \\ 0, & t \ge \pi. \end{cases}$$

Find the resulting motion.

2.22. Repeat Exercise 2.21 for $f(t) = [u(t) - u(t - \pi/k)] \sin kt$.

2.23. An EMF given by

$$e(t) = \begin{cases} \sin \omega t, & 0 \le t < \pi/\omega \\ 0, & t \ge \pi/\omega \end{cases}$$

is applied to a resistance and inductance in series. Find the expression for the current as a function of time, assuming that $i(0) = 0$.

2.24. The weight on a spring is pulled X_0 units beyond its equilibrium position and released. Find the resulting motion if no damping is present and the force on the spring is

(a) $u(t)t$

(b) $\begin{cases} t, & 0 \le t < \pi \\ 0, & t \ge \pi \end{cases}$

(Assume $m = 1$.)

2.25. A beam having one end clamped and the other free is called a *cantilever beam*. Placing the beam along the x axis, suppose it is clamped at $x = 0$ and free at $x = a$ and carries a uniform load F_0 per unit length. Find an expression for the deflection as a function of x.

2.26. Suppose the beam in Exercise 2.25 is clamped at both ends $x = 0$ and $x = a$. What is the deflection?

2.27. A cantilever beam is clamped at $x = 0$ and free at $x = a$. Find the deflection if the load is given by

$$F(x) = \begin{cases} F_0, & 0 \leq x < \tfrac{1}{2}a \\ 0, & x \geq \tfrac{1}{2}a. \end{cases}$$

2.28. Repeat Exercise 2.27 if

$$F(x) = \begin{cases} F_0 x, & 0 \leq x < \tfrac{1}{2}a \\ 0, & x \geq \tfrac{1}{2}a. \end{cases}$$

Exercises 2.29–2.40 involve unit impulse functions.

2.29. The weight on a spring is pulled b units beyond its equilibrium position and released. Find the resulting motion if no damping is present but an external force of $F_0 \delta(t - a)$ acts on the spring. The equation and boundary conditions are

$$mx'' + k^2 x = F_0 \delta(t - a), \quad x(0) = b, \quad x'(0) = 0.$$

2.30. Suppose the weight on the spring in Exercise 2.29 is initially at rest but acted on by a force equal to $A \sin \omega t$. At $t = 10$ the weight is struck from below by a blow with a strength of 2 units. Find the resulting motion.

2.31. In an *LRC* circuit $L = 0.5$ henry, $R = 10$ ohms, and $C = 2 \times 10^{-4}$ farad. Suppose a voltage impulse of strength 100 is applied to the circuit at $t = 1$ and $t = 2$. Find $q(t)$ if $q(0) = 0$ and $i(0) = 0$.

2.32. An inductor of L henrys and a capacitor of C farads are in series with a generator. Suppose a voltage impulse of strength E_0 is applied to the circuit at $t = 0$. Find the charge if $q(0) = 0$ and $i(0) = 0$.

2.33. Apply the impulse of Exercise 2.32 to an *LRC* circuit, with $i(0) = q(0) = 0$.

2.34. A beam is clamped at both ends $x = 0$ and $x = a$. Suppose a concentrated load P_0 acts vertically downward at $\frac{1}{4}a$. What is the resulting deflection?

2.35. A beam of length a is clamped at $x = 0$ and $x = a$ and carries concentrated loads P_1 and P_2 at $x = \frac{1}{3}a$ and $x = \frac{2}{3}a$, respectively. Find the resulting deflection.

2.36. A cantilever beam is clamped at $x = 0$ and free at $x = a$. Suppose the beam carries a uniform load F_0 on the left half of the beam and a concentrated load P_0 at $x = \frac{3}{4}a$. Find the deflection.

2.37. Suppose a beam is clamped at $x = 0$ and carries a concentrated load P_0 at the free end $x = a$; find the deflection.

2.38. What is the deflection of a cantilever beam fixed at $x = 0$ and free at $x = a$ if a concentrated load P_0 acts downward at $x = \frac{1}{2}a$?

2.39. If the ends of a beam are hinged, the boundary conditions become $Y = Y'' = 0$. Suppose that the ends are hinged at $x = 0$ and $x = a$ and that a concentrated load P_0 acts at $x = \frac{1}{3}a$. What is the deflection?

2.40. Consider the mechanical system and its electrical (force–voltage) analog shown in the diagram on page 74. It has two degrees of freedom. Recalling that $i = dq/dt$ and applying Kirchhoff's voltage law, we get

$$L_1 \frac{d^2q_1}{dt^2} + R \frac{dq_1}{dt} + \frac{1}{C_1} q_1 + \frac{1}{C_2} (q_1 - q_2) = e(t)$$

and

$$L_2 \frac{d^2q_2}{dt^2} + \frac{1}{C_2} (q_2 - q_1) = 0.$$

For the mechanical system the equations are

$$m_1 \frac{d^2x_1}{dt^2} + c \frac{dx_1}{dt} + k_1 x_1(t) + k_2[x_1(t) - x_2(t)] = \delta(t)$$

and

$$m_2 \frac{d^2x_2}{dt^2} + k_2[x_2(t) - x_1(t)] = 0.$$

Show that for the mechanical system

$$X_1(s) = \frac{m_2 s^2 + k_2}{(m_1 s^2 + cs + k_1 + k_2)(m_2 s^2 + k_2) - k_2^2},$$

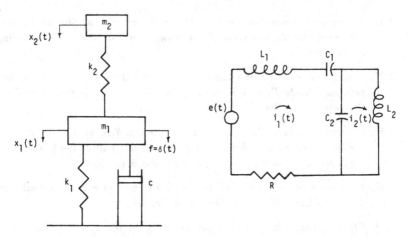

under suitable initial conditions. If $m_1 = m_2 = k_1 = k_2 = c = 1$, show that the zeros of the denominator have only negative real parts, using the Hurwitz criterion. What can be said about the resulting motion?

2.6. Transforms of Periodic Functions

A function is periodic with period T if, for all t, $f(t) = f(t + T)$. The following theorem enables us to find transforms of periodic functions.

Theorem 2.2. Suppose that $L\{f(t)\} = F(s)$ exists and that $f(t)$ is periodic with period T. Then

$$F(s) = \frac{1}{1 - e^{-sT}} \int_0^T e^{-st} f(t)\, dt \qquad (s > 0). \qquad (2.14)$$

Proof

$$F(s) = \int_0^\infty e^{-st} f(t)\, dt$$

$$= \int_0^T e^{-st} f(t)\, dt + \int_T^{2T} e^{-st} f(t)\, dt + \int_{2T}^{3T} e^{-st} f(t)\, dt + \cdots$$

$$= \sum_{n=0}^\infty \int_{nT}^{(n+1)T} e^{-st} f(t)\, dt.$$

Letting $u = t - nT$,

$$F(s) = \sum_{n=0}^{\infty} \int_0^T e^{-s(u+nT)} f(u + nT)\, du,$$

and by the periodicity,

$$F(s) = \sum_{n=0}^{\infty} \int_0^T e^{-su} f(u) e^{-nTs}\, du$$

$$= \sum_{n=0}^{\infty} e^{-nTs} \int_0^T e^{-su} f(u)\, du.$$

Now

$$\sum_{n=0}^{\infty} e^{-nTs} = 1 + e^{-Ts} + e^{-2Ts} + \cdots$$

is a geometric series (see Chapter 1) whose sum is $1/(1 - e^{-Ts})$. Hence

$$F(s) = \frac{1}{1 - e^{-Ts}} \int_0^T e^{-su} f(u)\, du.$$

Remark 2.2. We can see from equation (2.14) that to compute the transform of a periodic function, we need only the transform of the function identical to $f(t)$ over the first period and 0 elsewhere. [Being periodic, $f(t)$ is completely determined once its behavior over the first period is known.]

Example 2.12. Find the transform of the periodic function $f(t)$ of period 2 shown in Figure 2.18.

The function identical to $f(t)$ over the first period and 0 elsewhere is shown in Figure 2.19. By Example 2.6, its transform is

$$\frac{1}{s} - \frac{e^{-s}}{s^2} + \frac{e^{-2s}}{s^2}.$$

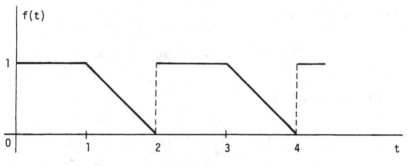

FIGURE 2.18

Consequently, by Theorem 2.2,

$$L\{f(t)\} = \frac{1}{1 - e^{-2s}} \left(\frac{1}{s} - \frac{e^{-s}}{s^2} + \frac{e^{-2s}}{s^2} \right).$$

Example 2.13. Find the Laplace transform of $f(t)$ shown in Figure 2.20.

The function identical to $f(t)$ over the first period and 0 elsewhere is shown in Figure 2.21; its transform is

$$\frac{1}{s^2} (1 - e^{-s})^2$$

by Example 2.7, so that

$$L\{f(t)\} = \frac{(1 - e^{-s})^2}{s^2(1 - e^{-2s})}$$

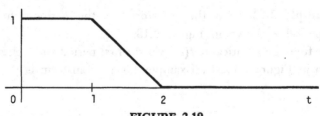

FIGURE 2.19

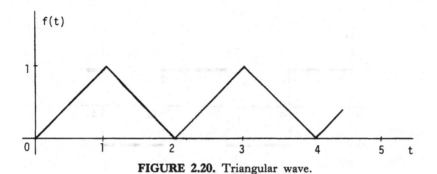

FIGURE 2.20. Triangular wave.

by Theorem 2.2. The last expression can be reduced to

$$L\{f(t)\} = \frac{(1 - e^{-s})^2}{s^2(1 - e^{-s})(1 + e^{-s})}$$

$$= \frac{1 - e^{-s}}{s^2(1 + e^{-s})}$$

$$= \frac{1}{s^2} \left(\frac{e^{s/2} - e^{-s/2}}{e^{s/2} + e^{-s/2}} \right)$$

$$= \frac{1}{s^2} \tanh \tfrac{1}{2}s.$$

As another illustration, consider the *square wave* in Figure 2.22. By Exercise 2.15(c), the transform of the function consisting of only the first period of $f(t)$ is

$$\frac{(1 - e^{-as})^2}{s}.$$

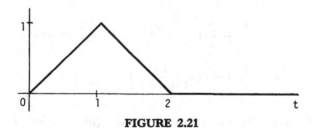

FIGURE 2.21

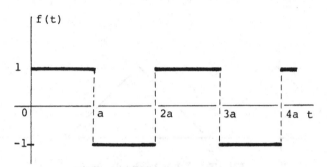

FIGURE 2.22. Square wave.

Since the period is $2a$, we obtain, by Theorem 2.2,

$$F(s) = \frac{(1 - e^{-as})^2}{s(1 - e^{-2as})},$$

which reduces to

$$F(s) = \frac{1 - e^{-as}}{s(1 + e^{-as})} = \frac{1}{s} \tanh \tfrac{1}{2}as.$$

Looking ahead to some of our applications, let us consider another method of finding the last transform. In terms of unit step functions,

$$f(t) = u(t) - 2u(t - a) + 2u(t - 2a) - \cdots \qquad (2.15)$$

and

$$L\{f(t)\} = \frac{1}{s} - \frac{2e^{-as}}{s} + \frac{2e^{-2as}}{s} - \frac{2e^{-3as}}{s} + \cdots \qquad (2.16)$$

$$= -\frac{1}{s} + \frac{2}{s} \left(1 - e^{-as} + e^{-2as} - e^{-3as} + \cdots \right)$$

$$= -\frac{1}{s} + \frac{2}{s} \left(\frac{1}{1 + e^{-as}}\right)$$

by the formula for the geometric series (see Section 1.3) with

$r = -e^{-as}$. Thus

$$L\{f(t)\} = \frac{1 - e^{-as}}{s(1 + e^{-as})} = \frac{1}{s} \tanh \tfrac{1}{2}as. \qquad (2.17)$$

We see, then, that Theorem 2.2 enables us to find transforms of periodic functions directly in closed form. On the other hand, the infinite series representation in equation (2.16) is perfectly valid and actually more convenient in some applications. One reason is that the calculation of the inverse transform of a closed expression may be far from simple, while the inverse transform of equation (2.16) can be found directly. [In this case, of course, the inverse of $(1/s)$ tanh $\tfrac{1}{2}as$ could be obtained from the table.] In fact, whenever possible, the simplest way of finding the inverse transform of a closed expression such as

$$F(s) = \frac{1 - e^{-s}}{s(1 - e^{-4s})}$$

is to write it as a series as follows:

$$\frac{1 - e^{-s}}{s(1 - e^{-4s})} = \frac{1}{s}(1 - e^{-s})(1 + e^{-4s} + e^{-8s} + e^{-12s} + \cdots)$$

$$= \frac{1}{s} - \frac{e^{-s}}{s} + \frac{e^{-4s}}{s} - \frac{e^{-5s}}{s} + \frac{e^{-8s}}{s} - \frac{e^{-9s}}{s} + \cdots.$$

Hence

$$f(t) = u(t) - u(t - 1) + u(t - 4) - u(t - 5) + u(t - 8) - \cdots \qquad (2.18)$$

(Figure 2.23). Unfortunately, this procedure cannot always be employed. In general, the inverse transform of a closed expression has to be found by the use of the complex inversion formula, a direct method of calculating inverse transforms. The same is true if a dif-

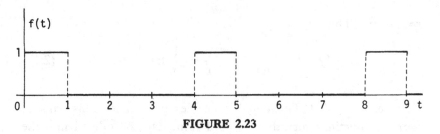

FIGURE 2.23

ferent form of the inverse is sought. Since the inversion formula requires complex variable theory, we shall postpone its discussion until Chapter 4.

As a final comment, the expression in equation (2.18) should not be viewed as an infinite series, since for any fixed t the number of terms is really finite. Indeed, $f(t)$ could have been written more simply in the form

$$f(t) = \begin{cases} 1, & 4n \le t < 4n + 1 \\ 0, & \text{otherwise.} \end{cases}$$

Similar remarks hold for equation (2.15) as well as other infinite-series expressions involving unit step functions.

Exercises

2.41. Use the exercises in Section 2.3 in conjunction with Theorem 2.2 to find the transforms of the periodic functions in the following diagrams:

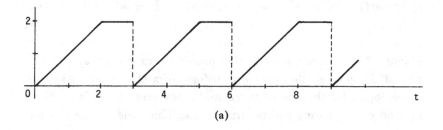

(a)

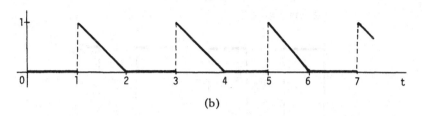

(b)

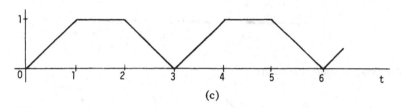

(c)

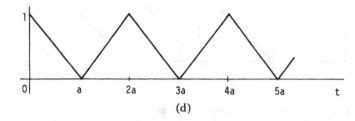

(d)

2.42. The series of impulses in the diagram may be represented by $\delta(t) +$ $\delta(t - a) + \delta(t - 2a) + \cdots$. Show that the transform is $1/(1 - e^{-as})$.

For the remaining exercises, determine the transform in two ways:

(a) Find $L\{f(t)\}$ by the use of Theorem 2.2.

(b) Write $f(t)$ in terms of unit step functions, and express $L\{f(t)\}$ in infinite-series form.

2.43. Referring to the diagram, note that $f(t)$ can be written

$$u(t) - u(t - a) + u(t - 2a) - u(t - 3a) + \cdots.$$

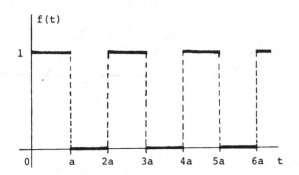

2.44. The *sawtooth* function:

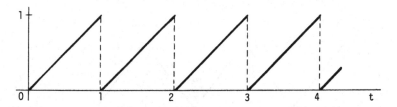

2.45. The half-wave rectification of the sine wave:

$$f(t) = \begin{cases} \sin \omega t, & 0 \leq t < \pi/\omega \\ 0, & \pi/\omega \leq t < 2\pi/\omega \\ f(t + 2\pi/\omega), & t \geq 2\pi/\omega. \end{cases}$$

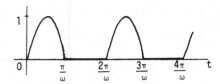

2.46. $f(t) = |\sin \omega t|$, the full-wave rectification of the sine wave:

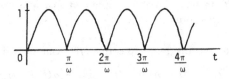

2.7. Applications

Example 2.14. Suppose an EMF is applied to a resistance and inductance in series. Given that $i(0) = 0$, find $i(t)$ if $e(t)$ is the half-wave rectification of $\sin \omega t$ (Figure 2.24):

$$e(t) = \begin{cases} \sin \omega t, & 0 \le t < \pi/\omega \\ 0, & \pi/\omega \le t < 2\pi/\omega \\ e(t + 2\pi/\omega), & t \ge 2\pi/\omega. \end{cases}$$

The equation is

$$L \frac{di(t)}{dt} + Ri(t) = u(t) \sin \omega t + u\left(t - \frac{\pi}{\omega}\right) \sin \omega\left(t - \frac{\pi}{\omega}\right)$$

$$+ u\left(t - \frac{2\pi}{\omega}\right) \sin \omega\left(t - \frac{2\pi}{\omega}\right) + \cdots,$$

and the transformed equation is

$$LsI(s) + RI(s) = \left(\frac{\omega}{s^2 + \omega^2}\right)(1 + e^{-\pi s/\omega} + e^{-2\pi s/\omega} + \cdots),$$

whence

$$I(s) = \frac{1}{Ls + R}\left(\frac{\omega}{s^2 + \omega^2}\right)(1 + e^{-\pi s/\omega} + e^{-2\pi s/\omega} + \cdots).$$

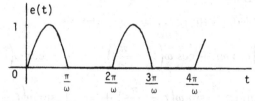

FIGURE 2.24. Half-wave rectification of $\sin \omega t$.

Now

$$\frac{1}{Ls + R}\left(\frac{\omega}{s^2 + \omega^2}\right)$$

$$= \frac{\omega/L}{[s + (R/L)](s^2 + \omega^2)}$$

$$= \frac{1}{R^2 + L^2\omega^2}\left[\frac{L\omega}{s + (R/L)} - \frac{L\omega s}{s^2 + \omega^2} + \frac{R\omega}{s^2 + \omega^2}\right].$$

Hence

$$i(t) = \frac{1}{R^2 + L^2\omega^2}\left\{u(t)(L\omega e^{-(R/L)t} - L\omega \cos \omega t + R \sin \omega t)\right.$$

$$+ u\left(t - \frac{\pi}{\omega}\right)\left[L\omega e^{-(R/L)[t-(\pi/\omega)]} - L\omega \cos \omega\left(t - \frac{\pi}{\omega}\right)\right.$$

$$\left. + R \sin \omega\left(t - \frac{\pi}{\omega}\right)\right]$$

$$+ u\left(t - \frac{2\pi}{\omega}\right)\left[L\omega e^{-(R/L)[t-(2\pi/\omega)]} - L\omega \cos \omega\left(t - \frac{2\pi}{\omega}\right)\right.$$

$$\left.\left. + R \sin \omega\left(t - \frac{2\pi}{\omega}\right)\right] + \cdots\right\}.$$

Given any interval, there are only a finite number of terms. Suppose we obtain an expression for the current on the interval $n\pi/\omega \leq t < (n + 1)(\pi/\omega)$. Then

$$i(t) = \frac{1}{R^2 + L^2\omega^2}\left\{L\omega(e^{-(R/L)t} + e^{-(R/L)[t-(\pi/\omega)]}\right.$$

$$+ e^{-(R/L)[t-(2\pi/\omega)]} + \cdots + e^{-(R/L)[t-(n\pi/\omega)]})$$

$$- L\omega\left[\cos \omega t + \cos \omega\left(t - \frac{\pi}{\omega}\right) + \cdots + \cos \omega\left(t - \frac{n\pi}{\omega}\right)\right]$$

$$\left. + R\left[\sin \omega t + \sin \omega\left(t - \frac{\pi}{\omega}\right) + \cdots + \sin \omega\left(t - \frac{n\pi}{\omega}\right)\right]\right\}.$$

This expression can be simplified further since

$$\cos \omega \left(t - \frac{n\pi}{\omega} \right) = (-1)^n \cos \omega t$$

and

$$\sin \omega \left(t - \frac{n\pi}{\omega} \right) = (-1)^n \sin \omega t,$$

while

$$e^{-(R/L)t} + e^{-(R/L)[t-(\pi/\omega)]} + \cdots + e^{-(R/L)[t-(n\pi/\omega)]}$$

$$= e^{-(R/L)t}(1 + e^{R\pi/L\omega} + \cdots + e^{nR\pi/L\omega})$$

$$= \frac{e^{-(R/L)t}(1 - e^{[(n+1)R\pi]/L\omega})}{1 - e^{R\pi/L\omega}}.$$

[Here we have used the formula $1 + x + x^2 + \cdots + x^n = (1 - x^{n+1})/(1 - x)$.] Thus

$$i(t) = \begin{cases} \dfrac{1}{R^2 + L^2\omega^2} \left[\dfrac{L\omega e^{-(R/L)t}(1 - e^{[(n+1)R\pi]/L\omega})}{1 - e^{R\pi/L\omega}} \right], & \text{for } n \text{ odd} \\[2em] \dfrac{1}{R^2 + L^2\omega^2} \left[\dfrac{L\omega e^{-(R/L)t}(1 - e^{[(n+1)R\pi]/L\omega})}{1 - e^{R\pi/L\omega}} \right. \\[1.5em] \qquad\qquad \left. - L\omega \cos \omega t + R \sin \omega t \right], & \text{for } n \text{ even;} \end{cases}$$

$n\pi/\omega \le t < [(n + 1)\pi]/\omega, \; n = 0, 1, 2, \ldots$.

As a final task, suppose we examine the steady state. Observe that for large n

$$e^{[(n+1)R\pi]/L\omega} \gg 1$$

(very much larger than 1), so that

$$\frac{e^{-(R/L)t}(1 - e^{[(n+1)R\pi]/L\omega})}{1 - e^{R\pi/L\omega}}$$

reduces to

$$\frac{e^{-(R/L)t}\left(-e^{[(n+1)R\pi]/L\omega}\right)}{1-e^{R\pi/L\omega}}=-\frac{e^{-(R/L)t}e^{nR\pi/L\omega}e^{R\pi/L\omega}}{1-e^{R\pi/L\omega}}$$

$$=\frac{e^{[-R(\omega t-n\pi)]/L\omega}e^{R\pi/L\omega}}{e^{R\pi/L\omega}-1}=\frac{e^{-[R(\omega t-n\pi)]/L\omega}}{1-e^{-R\pi/L\omega}}.$$

Since $n\pi/\omega \le t < [(n+1)\pi]/\omega$, $\omega t - n\pi \ge 0$, and we see that $i(t)$ is a decaying exponential for n odd, that is, whenever the input voltage is 0.

Example 2.15. Suppose the input signal $e_1(t)$ in Figure 2.25 is applied to the circuit in Figure 2.26. Let us compute the output voltage $e_2(t)$. The equation is

$$L\frac{di(t)}{dt}+Ri(t)=E_1[u(t)-u(t-t_0)+u(t-a)-u(t-a-t_0)$$
$$+u(t-2a)-u(t-2a-t_0)+\cdots],$$

and the transformed equation is

$$LsI(s)+RI(s)=E_1\bigg(\frac{1}{s}-\frac{e^{-t_0s}}{s}+\frac{e^{-as}}{s}-\frac{e^{-(a+t_0)s}}{s}$$
$$+\frac{e^{-2as}}{s}-\frac{e^{-(2a+t_0)s}}{s}+\cdots\bigg),$$

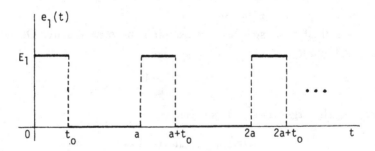

FIGURE 2.25. Input signal consisting of a chain of pulses.

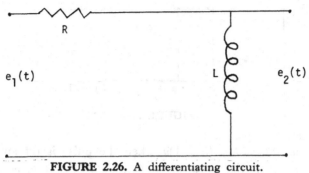

FIGURE 2.26. A differentiating circuit.

whence

$$I(s) = \frac{E_1}{Ls + R}\left(\frac{1}{s} - \frac{e^{-t_0 s}}{s} + \frac{e^{-as}}{s} - \frac{e^{-(a+t_0)s}}{s} + \frac{e^{-2as}}{s}\right.$$

$$\left. - \frac{e^{-(2a+t_0)s}}{s} + \cdots\right).$$

Since the voltage drop across an inductor is given by $L[di(t)/dt]$, we get (under zero initial conditions)

$$E_2(s) = LsI(s) = \frac{LE_1 s}{Ls + R}\left(\frac{1}{s} - \frac{e^{-t_0 s}}{s} + \frac{e^{-as}}{s} - \frac{e^{-(a+t_0)s}}{s}\right.$$

$$\left. + \frac{e^{-2as}}{s} - \frac{e^{-(2a+t_0)s}}{s} + \cdots\right)$$

$$= \frac{E_1}{s + (R/L)}\left(1 - e^{-t_0 s} + e^{-as} - e^{-(a+t_0)s}\right.$$

$$\left. + e^{-2as} - e^{-(2a+t_0)s} + \cdots\right).$$

Hence

$$e_2(t) = E_1[e^{-(R/L)t} - u(t - t_0)e^{-(R/L)(t-t_0)} + u(t - a)e^{-(R/L)(t-a)}$$

$$- u(t - a - t_0)e^{-(R/L)(t-a-t_0)} + u(t - 2a)e^{-(R/L)(t-2a)}$$

$$- u(t - 2a - t_0)e^{-(R/L)(t-2a-t_0)} + \cdots]. \qquad (2.19)$$

To see how this result can be interpreted, let us find $e_2(t)$ on the

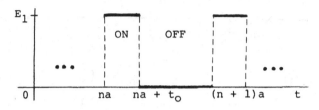

FIGURE 2.27

intervals $na + t_0 \leq t < (n + 1)a$ when the pulse is off and $na \leq t < na + t_0$ when the pulse is on (Figure 2.27). It is clear from equation (2.19) that in the former case we get

$$e_2(t) = E_1(e^{-(R/L)t} - e^{-(R/L)(t-t_0)} + e^{-(R/L)(t-a)} - e^{-(R/L)(t-a-t_0)}$$
$$+ \cdots + e^{-(R/L)(t-na)} - e^{-(R/L)(t-na-t_0)}), \qquad (2.20)$$

while in the latter case the last term of the sum is missing. Consequently, if the pulse is off, equation (2.20) can be written

$$e_2(t) = E_1(e^{-Rt/L} - e^{-Rt/L}e^{Rt_0/L} + e^{-Rt/L}e^{Ra/L} - e^{-Rt/L}e^{Ra/L}e^{Rt_0/L}$$
$$+ \cdots + e^{-Rt/L}e^{nRa/L} - e^{-Rt/L}e^{nRa/L}e^{Rt_0/L})$$
$$= E_1[e^{-Rt/L}(1 + e^{Ra/L} + \cdots + e^{nRa/L})$$
$$- e^{-Rt/L}e^{Rt_0/L}(1 + e^{Ra/L} + \cdots + e^{nRa/L})]$$
$$= E_1e^{-Rt/L}\left[\frac{1 - e^{[(n+1)Ra]/L}}{1 - e^{Ra/L}} - e^{Rt_0/L}\left(\frac{1 - e^{[(n+1)Ra]/L}}{1 - e^{Ra/L}}\right)\right]; \quad (2.21)$$

$na + t_0 \leq t < (n + 1)a$, $n = 0, 1, 2, \ldots$. Similarly, when the pulse is on,

$$e_2(t) = [e^{-Rt/L}(1 + e^{Ra/L} + \cdots + e^{nRa/L})$$
$$- e^{-Rt/L}e^{Rt_0/L}(1 + e^{Ra/L} + \cdots + e^{[(n-1)Ra]/L})]$$
$$= E_1e^{-Rt/L}\left[\frac{1 - e^{[(n+1)Ra]/L}}{1 - e^{Ra/L}} - e^{Rt_0/L}\left(\frac{1 - e^{nRa/L}}{1 - e^{Ra/L}}\right)\right]; \quad (2.22)$$

$na \leq t < na + t_0$, $n = 0, 1, 2, \ldots$.

To obtain the steady-state solution we observe as before that

$$e^{nRa/L} \gg 1$$

for large n. Consequently equation (2.21) reduces to

$$e_2(t) = E_1 e^{-Rt/L}\left[\frac{-e^{[(n+1)Ra]/L}}{1 - e^{Ra/L}} - e^{Rt_0/L}\left(\frac{-e^{[(n+1)Ra]/L}}{1 - e^{Ra/L}}\right)\right]$$

$$= -E_1 e^{-Rt/L} e^{Rt_0/L}\left(\frac{e^{-Rt_0/L}e^{[(n+1)Ra]/L}}{1 - e^{Ra/L}} - \frac{e^{[(n+1)Ra]/L}}{1 - e^{Ra/L}}\right)$$

$$= -E_1 e^{[-R(t-na-t_0)]/L}\left[\frac{e^{Ra/L}\left(e^{-Rt_0/L} - 1\right)}{1 - e^{Ra/L}}\right]$$

$$= -E_1 e^{-(R/L)(t-na-t_0)}\left(\frac{e^{-Rt_0/L} - 1}{e^{-Ra/L} - 1}\right).$$

Similarly, equation (2.22) becomes

$$e_2(t) = E_1 e^{-Rt/L}\left[\frac{-e^{[(n+1)Ra]/L}}{1 - e^{Ra/L}} - e^{Rt_0/L}\left(\frac{-e^{nRa/L}}{1 - e^{Ra/L}}\right)\right]$$

$$= E_1 e^{[-R(t-na)]/L}\left(\frac{-e^{Ra/L} + e^{Rt_0/L}}{1 - e^{Ra/L}}\right)$$

$$= E_1 e^{-(R/L)(t-na)}\left(\frac{1 - e^{-(R/L)(a-t_0)}}{1 - e^{-Ra/L}}\right).$$

Thus we have obtained the steady-state solution. If, in addition, $R \gg L$, the circuit is called a *differentiating circuit*, whose output is shown in Figure 2.28. Such a circuit is useful in locating edges of pulses; it is, in fact, one of the fundamental circuits in electronics, although it is not usually used in this simple form but with another active element, such as a transistor. For a differentiating circuit, we have, ideally,

$$\frac{de_1(t)}{dt} = e_2(t).$$

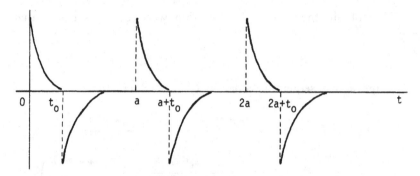

FIGURE 2.28. Effect of a differentiating circuit on a chain of pulses.

Since

$$\frac{d}{dt}\,u(t) = \delta(t),$$

we see from Figure 2.28 that this behavior is closely approximated if $R \gg L$.

Exercises

2.47. An object having mass m is initially at rest and receives a blow, or impulse, of strength P at $t = 0$, $t = a$, $t = 2a$, etc. The equation may be written

$$m\,\frac{dv}{dt} = P[\delta(t) + \delta(t - a) + \delta(t - 2a) + \cdots].$$

Find the velocity as a function of time.

2.48. A voltage impulse of strength E_0 is applied to an LC circuit at regular intervals, say $t = 0$, a, $2a$, etc. Find an expression for the charge $q(t)$ assuming that $q(0) = 0$ and $i(0) = 0$.

2.49. In an LRC circuit, $L = 0.5$ henry, $R = 10$ ohms, and $C = 2 \times 10^{-4}$ farad. Suppose a voltage impulse of strength 100 is applied to the circuit at $t = 0$, $t = 1$, $t = 2$, etc. Find $q(t)$ if $q(0) = i(0) = 0$.

2.50. A voltage impulse is applied to an LR circuit with initial current 0. As before, the strength of the impulse is E_0, applied at $t = 0$, a, $2a$, etc.

    (a)   Find the current.

    (b)   Find the steady-state expression for the current.

2.51. Repeat Exercise 2.50(a) assuming that the initial current is I_0.

2.52. Suppose the sequence of impulses in Exercise 2.50 is applied to a general LRC circuit with zero initial conditions. Find $q(t)$.

2.53. The sequence of pulses in the diagram may be regarded as a sequence of dots in the Morse telegraph code. If applied to an LR circuit for which $i(0) = 0$, find

    (a) $i(t)$.

    (b)   The steady state for the current.

    (c)   The output voltage $L(di/dt)$ from part (b).

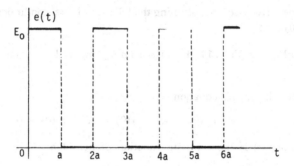

2.54. Repeat Exercise 2.53(a) using the square-wave input shown in the following diagram.

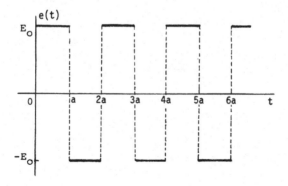

2.55. Suppose the weight of mass 1 on a spring is initially at rest but acted on by the force pictured in the following diagram. Find the resulting motion if no damping is present.

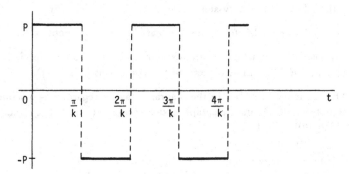

2.56. Repeat Exercise 2.55 assuming the following initial conditions: $x(0) = 0$, $x'(0) = X_1$.

2.57. Work Example 2.14 if $e(t) = |\sin t|$, the full-wave rectification of $\sin t$.

2.58. Solve the spring equation

$$x'' + k^2 x = f, \qquad x(0) = X_0, \quad x'(0) = 0,$$

where $f(t)$ is the sawtooth function of period π.

3

Sketch of Complex
Variable Theory

3.1. Basic Concepts

In this section we are going to consider briefly some of the basic concepts from complex variable theory. We shall need just enough to obtain a method for finding inverse transforms, which means that the reader will be referred to Appendix A for many of the details.

Let $i = \sqrt{-1}$ or, equivalently, $i^2 = -1$. We shall consider a complex number as having the form $a + bi$, where a and b are real numbers called the *real* and *imaginary parts*, respectively. If $z = a + bi$, then the *complex conjugate*, denoted by $\bar{z}$, is defined to be $a - bi$.

Complex numbers have been given a geometric interpretation, as illustrated in Figure 3.1. Thus $x + iy$ can be represented as a point with coordinates (x, y) or as a vector from the origin to (x, y). The length of the arrow, denoted by r in the figure, is called the *modulus* or *absolute value*. So if $z = x + iy$, the modulus is found to be

$$| z | = | x + iy | = (x^2 + y^2)^{1/2} = r.$$

The angle in Figure 3.1 is called the *argument* of the complex number.

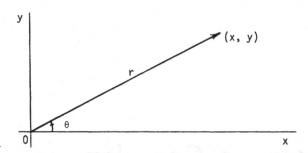

FIGURE 3.1. Geometric interpretation of a complex number.

Clearly,

$$x + iy = r(\cos \theta + i \sin \theta),$$

usually called the *polar form*.

Our next idea is the familiar notion of a function. If for every value of the variable z there corresponds a value of the variable w, we call w a function of z, denoted by $w = f(z)$. If for each value of z there corresponds only one value of w, the function is said to be *single-valued*. And just as in the case of functions of real variables, we define the derivative of a single-valued function $f(z)$ by

$$f'(z) = \lim_{\Delta z \to 0} \frac{f(z + \Delta z) - f(z)}{\Delta z}$$

provided that the limit exists and is independent of the manner in which the complex increment Δz goes to 0. If a function is differentiable in a region D, it is said to be *analytic* in D. A necessary condition that $f(z)$ be analytic is given in Appendix A, Section 1. It is also shown that if

$$f(z) = u(x, y) + iv(x, y),$$

$$f'(z) = \frac{\partial u}{\partial x} + i\frac{\partial v}{\partial x} = \frac{\partial v}{\partial y} - i\frac{\partial u}{\partial y},$$

from which follow the usual rules for differentiation familiar from calculus. For example, if $f(z) = z^2$, then

$$f(z) = (x + iy)^2 = x^2 - y^2 + 2xyi$$

and

$$f'(z) = \frac{\partial v}{\partial y} - i\frac{\partial u}{\partial y}$$

$$= 2x + 2yi$$

$$= 2z.$$

Thus $(d/dz)z^2 = 2z$.

Exponential, trigonometric, and other transcendental functions can be defined in terms of power series. For example,

$$e^z = 1 + z + \frac{z^2}{2!} + \frac{z^3}{3!} + \cdots, \tag{3.1}$$

$$\sin z = z - \frac{z^3}{3!} + \frac{z^5}{5!} - \cdots, \tag{3.2}$$

$$\cos z = 1 - \frac{z^2}{2!} + \frac{z^4}{4!} - \cdots. \tag{3.3}$$

As usual,

$$\cosh z = \frac{e^z + e^{-z}}{2} \quad \text{and} \quad \sinh z = \frac{e^z - e^{-z}}{2}.$$

Formally replacing z by $i\theta$ (θ real) in equation (3.1) and comparing the series to equations (3.2) and (3.3) suggests the definition

$$e^{i\theta} = \cos\theta + i\sin\theta, \tag{3.4}$$

known as *Euler's identity*, which will be used frequently in later sections. Other useful relationships are

$$\sin \theta = \frac{e^{i\theta} - e^{-i\theta}}{2i}, \tag{3.5}$$

$$\cos \theta = \frac{e^{i\theta} + e^{-i\theta}}{2}, \tag{3.6}$$

whence

$$\sinh i\theta = i \sin \theta \tag{3.7}$$

and

$$\cosh i\theta = \cos \theta. \tag{3.8}$$

Relationships (3.5) and (3.6) can be obtained from Euler's identity or from the above series.

Having briefly discussed the derivative, it seems natural to consider the integral next. This concept is somewhat more complicated; in fact, we must first consider what is meant by a *path* in the complex plane. The simplest way to define a path is as a curve in the plane described by a set of two parametric equations. For example, we know from analytic geometry that

$$\left. \begin{array}{l} x(t) = t \\ y(t) = t^2 \end{array} \right\} \quad -\infty < t < \infty$$

describes the parabola $y = x^2$. Clearly, the set $t + it^2$ (all t) represents the same curve in the complex plane. Similarly, $e^{i\theta} = \cos \theta + i \sin \theta$, $0 \leq \theta < 2\pi$, is the unit circle. We see, then, that a path $z(t)$ can be defined as a function from an interval $a \leq t \leq b$ into the complex plane. If $z(a) = z(b)$, the path is said to be closed, and if $z(t_1) \neq z(t_2)$ for any t_1 and t_2, the path is called simple, having no loops. The curve in Figure 3.2 is an example of a simple closed path. Furthermore, one usually assigns a direction to the path, say the direction of

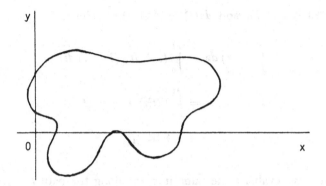

FIGURE 3.2. A simple closed path.

increasing t. In most applications it is also assumed that $z(t)$ has a sectionally continuous derivative. So if $f(z)$ is continuous, then $f(z(t))$ is also continuous, and we define

$$\int_C f(z)\, dz = \int_a^b f(z(t))\, dz(t),$$

where C is the path $z(t)$, $a \le t \le b$.

Example 3.1. Suppose that $f(z) = z^2$ and that the path is a straight line segment from 0 to $1 + i$ (Figure 3.3). Then $z(t) =$

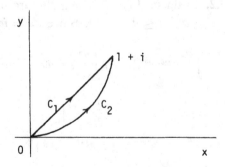

FIGURE 3.3. Two different paths of integration between 0 and $1 + i$.

$t + it$ $(0 \leq t \leq 1)$, and $dz(t) = (1 + i) dt$. Hence

$$\int_{C_1} f(z) \, dz = \int_0^1 (t + it)^2 (1 + i) \, dt$$

$$= \int_0^1 (2t^2 i)(1 + i) \, dt$$

$$= -\tfrac{2}{3} + \tfrac{2}{3}i. \tag{3.9}$$

Now we evaluate the same integral along the path $t + it^2$ (Figure 3.3):

$$\int_{C_2} f(z) \, dz = \int_0^1 (t + it^2)^2 (1 + 2ti) \, dt$$

$$= \int_0^1 (t^2 - t^4 + 2t^3 i)(1 + 2ti) \, dt$$

$$= -\tfrac{2}{3} + \tfrac{2}{3}i.$$

We note that the two results are the same. In fact, if a function is analytic in a region, the integral depends only on the end points, not on the particular path chosen. (For a proof, see Appendix A, Section 2.)

Example 3.2. Evaluate $\int_C z^2 \, dz$ along the arc of the circle in Figure 3.4, i.e., $z = e^{i\theta}$ $(0 \leq \theta \leq \pi/2)$. Since $dz = ie^{i\theta} \, d\theta$, we have

$$\int_0^{\pi/2} e^{2i\theta} ie^{i\theta} \, d\theta = i \int_0^{\pi/2} e^{3i\theta} \, d\theta$$

$$= \frac{i}{3i} e^{3i\theta} \Big|_0^{\pi/2}$$

$$= \tfrac{1}{3}(e^{3\pi i/2} - 1)$$

$$= -\tfrac{1}{3} - \tfrac{1}{3}i.$$

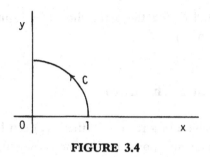

FIGURE 3.4

We have used the fact that

$$e^{3\pi i/2} = \cos\frac{3\pi}{2} + i\sin\frac{3\pi}{2}$$

by Euler's identity (3.4).

As a final observation, suppose that C encloses a region that is not simply connected, meaning that it contains a finite number of *holes*, as in Figure 3.5. Then if $f(z)$ is analytic in the shaded region,

$$\int_C f(z)\,dz = \int_{C_1} f(z)\,dz + \int_{C_2} f(z)\,dz + \int_{C_3} f(z)\,dz, \qquad (3.10)$$

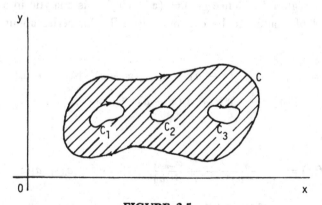

FIGURE 3.5

where C_1, C_2, and C_3 are the paths shown in Figure 3.5. (See Appendix A, Section 3.)

3.2. The Residue Theorem

If $f(z)$ is analytic in a region D, then $f(z)$ can be expanded in a Taylor series about any point in D. For certain functions that are not analytic in the entire region we shall obtain a generalization of the Taylor series, called the *Laurent series*.

To see how a function may fail to be analytic, consider $f(z) = g(z)/(z - a)$, where a is in D and $g(z)$ is analytic in D. The point $z = a$ is called an *isolated singularity* since $f(z)$ fails to be analytic there but is analytic in every neighborhood of a; the same is true of $g(z)/(z - a)^n$. To distinguish between the two cases, the former is called a *simple pole* and the latter a *pole of order n*. The function $f(z) = (\sin z)/z$ has a singularity at the origin, but since $\lim_{z\to 0} (\sin z)/z = 1$, it is called a *removable singularity*. Another type of singularity, called a *branch point*, will not be encountered in the remainder of the chapter. (See Appendix A, Section 4.)

Suppose we now consider a function $f(z)$ having a pole of order n at $z = a$; then, by definition, $f(z) = g(z)/(z - a)^n$ for $g(z)$ analytic in some region D. Hence $g(z) = (z - a)^n f(z)$ is analytic in a neighborhood of a and can be expanded in a Taylor series about $z = a$, that is,

$$g(z) = a_{-n} + a_{-n+1}(z - a) + a_{-n+2}(z - a)^2 + \cdots + a_0(z - a)^n$$
$$+ a_1(z - a)^{n+1} + a_2(z - a)^{n+2} + \cdots ,$$

whence

$$f(z) = \frac{a_{-n}}{(z - a)^n} + \frac{a_{-n+1}}{(z + a)^{n-1}} + \cdots + a_0 + a_1(z - a)$$
$$+ a_2(z - a)^2 + \cdots .$$

The last series is called a *Laurent series* for $f(z)$. The form of the series is revealing: The order of the pole corresponds to the index $-n$; if the function is analytic in D or if the singularity is removable, there are no negative indices, and the series reduces to a Taylor series.

As an example of a Laurent series, suppose we expand e^z/z about the origin, using the Maclaurin series of e^z. Since

$$e^z = 1 + z + \frac{z^2}{2!} + \frac{z^3}{3!} + \cdots ,$$

$$\frac{e^z}{z} = \frac{1}{z} + 1 + \frac{z}{2!} + \frac{z^2}{3!} + \cdots .$$

The presence of the term $1/z$ confirms the fact that the function has a pole of order 1 at $z = 0$.

As another example, consider

$$f(z) = \frac{1}{z(z - i)^2(z + 3)^4}.$$

Clearly, $f(z)$ has a pole of order 1 at the origin, a pole of order 2 at $z = i$, and a pole of order 4 at $z = -3$. The form of the corresponding Laurent series leads to the same conclusion. For example,

$$f(z) = \frac{A}{z + 3} + \frac{B}{(z + 3)^2} + \frac{C}{(z + 3)^3} + \frac{D}{(z + 3)^4} + \text{other terms};$$

the existence of the coefficient a_{-4} shows that $f(z)$ has a pole of order 4 at $z = -3$.

If a series has an infinite number of negative indices, we have an *essential singularity*. For example,

$$\sin \frac{1}{z} = \frac{1}{z} - \frac{1}{3!\, z^3} + \frac{1}{5!\, z^5} - \cdots$$

has an essential singularity at $z = 0$.

Next, we consider the case where $f(z)$ is analytic in a region bounded by a simple closed curve C except for a pole of order n at $z = a$, inside C. It is shown in Appendix A, Section 5, that

$$\int_C \frac{dz}{(z-a)^n} = \begin{cases} 0, & n \neq 1 \\ 2\pi i, & n = 1. \end{cases} \tag{3.11}$$

Integrating the series

$$f(z) = \frac{a_{-n}}{(z-a)^n} + \frac{a_{-n+1}}{(z-a)^{n-1}} + \cdots + \frac{a_{-1}}{z-a} + a_0$$
$$+ a_1(z-a) + \cdots$$

termwise about C, we get

$$\int_C f(z)\,dz = 2\pi i a_{-1}.$$

This remarkable formula provides the key to evaluating integrals around a simple closed path. Since the coefficient a_{-1} appears in the formula, it has been given a special name, the *residue* of $f(z)$ at the pole $z = a$. We see, then, that the evaluation of $\int_C f(z)\,dz$ is reduced to finding the residue a_{-1}. In fact, because of equation (3.10), a more general statement can be made:

Theorem 3.1. If $f(z)$ is analytic within and on a simple closed curve C (of finite length) except for a finite number of poles inside C, then

$$\int_C f(z)\,dz = 2\pi i S,$$

where S is the sum of the residues at the different poles.

We can find these residues, one at a time, by the formula

$$a_{-1} = \lim_{z \to a} \frac{1}{(n-1)!} \frac{d^{n-1}}{dz^{n-1}} [(z-a)^n f(z)], \tag{3.12}$$

where n is the order of the pole at $z = a$. (See Appendix A, Section 6.) If $n = 1$, equation (3.12) reduces to

$$a_{-1} = \lim_{z \to a} (z - a) f(z). \tag{3.13}$$

If $f(z)$ is a rational function, the procedure is particularly simple. For example, if $f(z) = 1/z(z - 1)^3$,

$$(z - 1)^3 f(z) = \frac{1}{z}.$$

Hence the residue at $z = 1$, where $f(z)$ has a pole of order 3, is

$$\frac{1}{2!} \frac{d^2}{dz^2} (z - 1)^3 f(z) = \frac{1}{2} \frac{d^2}{dz^2} \frac{1}{z} \bigg|_{z=1} = 1.$$

Remark 3.1. For completeness we note that if C is a simple closed curve containing $z = 1$ but not $z = 0$, the other singularity, then

$$\int_C \frac{dz}{z(z - 1)^3} = 2\pi i.$$

4

The Complex Inversion Formula

4.1. The Variable s

If s is a complex variable,

$$s = x + iy,$$

and the Laplace transform of the real-valued function f becomes

$$F(s) = \int_0^\infty e^{-st} f(t)\, dt \tag{4.1}$$

$$= \int_0^\infty e^{-xt} e^{-iyt} f(t)\, dt$$

$$= \int_0^\infty e^{-xt} \cos yt\, f(t)\, dt - i \int_0^\infty e^{-xt} \sin yt\, f(t)\, dt.$$

Since

$$|\, e^{-xt} \cos yt\, f(t)\, | \leq |\, e^{-st} f(t)\, |$$

and

$$| e^{-xt} \sin yt\, f(t) | \leq | e^{-st} f(t) |,$$

the integral (4.1) exists for $x > \alpha$ if f is of exponential order $e^{\alpha t}$ and sectionally continuous (Theorem 1.3). Writing

$$u(x, y) = \int_0^\infty e^{-xt} \cos yt\, f(t)\, dt$$

and

$$v(x, y) = - \int_0^\infty e^{-xt} \sin yt\, f(t)\, dt,$$

we obtain

$$F(s) = u(x, y) + iv(x, y);$$

we can easily check that the partial derivatives of u and v are continuous and that the Cauchy–Riemann conditions are satisfied (see Appendix A, Section 1). Consequently, $F(s)$ is analytic in the half-plane $x > \alpha$, where $s = x + iy$.

In making the transition from real to complex s, one more observation is needed. When calculating the transform (4.1) for $s = x$, as in Chapter 1, we obtain a real function $G(x)$ identical to $F(s)$ to the right of α along the real axis. Now if $s = x + iy$ and $G(s)$ is analytic in the half-plane $x > \alpha$, $F(s) \equiv G(s)$, since two different analytic functions cannot be identical on a line. More precisely, if $\{s_n\}$ is a sequence in the half-plane converging to some point in the half-plane and if $F(s_n) = G(s_n)$ for all n, then $F(s) \equiv G(s)$. (For a proof, see any standard text on complex variable theory.) Consequently, the integrations can be carried out as if s were real. Moreover, all our transforms in Appendix B are valid when s is complex, as are the properties of the transform discussed in the first two chapters (with the exception of Theorem 1.7, where s is assumed real).

4.2. The Inversion Integral

We are now ready to apply the theory of residues to the process of finding inverse transforms. If

$$L\{f(t)\} = F(s) = \int_0^\infty e^{-st} f(t)\, dt \qquad [\mathrm{Re}(s) > \alpha],$$

the inverse transform is given by

$$f(t) = \frac{1}{2\pi i} \int_{c-i\infty}^{c+i\infty} e^{st} F(s)\, ds \qquad (t > 0), \tag{4.2}$$

and $f(t) = 0$ for $t < 0$. The limits of integration are understood to mean

$$\lim_{b\to\infty} \int_{c-ib}^{c+ib} e^{st} F(s)\, ds;$$

in other words, the path of integration is a line parallel to the imaginary axis. (For a motivation of this definition, see Appendix A, Section 7.) It is also understood that the line lies to the right of all singularities; such a line always exists, since $F(s)$ is analytic in a half-plane $x > \alpha$.

In view of our discussion of residues it is not clear how such an integration can be performed. However, suppose that the only singularities are poles, finite in number. We may then choose some vertical line to the right of all singularities and form the closed curve in Figure 4.1. The radius R of the circular portion is so large that all the poles lie inside the curve. The integral around the entire curve can then be evaluated by the residue theorem. Now if $|F(s)| < MR^{-k}$ for M and k positive constants, it can be shown that as $R \to \infty$ the integral (4.2) will approach 0 along the circular portion of the curve. This implies that the integral around the entire curve approaches the integral along the line in the limit. Fortunately, the

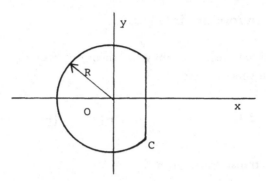

FIGURE 4.1. The basic path of integration of the inversion integral (before $R \to \infty$).

condition is satisfied for most transforms we normally consider. Thus the main result of this section can be stated as follows:

Theorem 4.1. $f(t)$ equals the sum of the residues of $e^{st}F(s)$ at the poles of $F(s)$.

To illustrate the procedure, suppose we find the inverse transforms of some known functions by the use of residues.

Example 4.1

$$L^{-1}\left\{\frac{1}{s-a}\right\} = \lim_{s \to a} (s-a)\,\frac{e^{st}}{s-a} = e^{at}$$

by formula (3.13).

Example 4.2. Find

$$L^{-1}\left\{\frac{1}{s(s-1)^2}\right\}.$$

Since there are two poles, the residues must be computed separately. For the residue at $s = 0$

$$\lim_{s \to 0} s \frac{e^{st}}{s(s-1)^2} = 1.$$

At $s = 1$ we have a pole of order 2 so that the residue can be obtained by formula (3.12). Thus

$$\frac{1}{1!} \lim_{s \to 1} \frac{d}{ds} (s-1)^2 \frac{e^{st}}{s(s-1)^2} = \lim_{s \to 1} \frac{d}{ds} \frac{e^{st}}{s}$$

$$= \lim_{s \to 1} \frac{st e^{st} - e^{st}}{s^2}$$

$$= te^t - e^t,$$

and

$$L^{-1}\left\{\frac{1}{s(s-1)^2}\right\} = 1 - e^t + te^t.$$

Example 4.3. Consider

$$F(s) = \frac{1}{s(s^2+1)} = \frac{1}{s(s-i)(s+i)},$$

which has simple poles at $s = 0$, i, and $-i$. At $s = 0$ we get

$$\lim_{s \to 0} \frac{s e^{st}}{s(s^2+1)} = 1.$$

At $s = -1$,

$$\lim_{s \to -i} \frac{(s+i)e^{st}}{s(s-i)(s+i)} = \lim_{s \to -i} \frac{e^{st}}{s(s-i)}$$

$$= \frac{e^{-it}}{-i(-2i)} = \frac{e^{-it}}{-2},$$

and at $s = i$,

$$\lim_{s \to i} \frac{e^{st}}{s(s+i)} = \frac{e^{it}}{-2}.$$

Hence

$$f(t) = 1 - \frac{e^{it} + e^{-it}}{2} = 1 - \cos t,$$

by equation (3.6).

We can see from these examples that the complex inversion formula provides a good alternative to the method of partial fractions.

Exercises

4.1. Use the complex inversion formula to find the inverse transforms of the functions in Exercises 1.20–1.31 (Chapter 1).

5

Convolutions

5.1. The Convolution Theorem

Suppose for two functions f and g we have that $L\{f(t)\} = F(s)$ and $L\{g(t)\} = G(s)$. Now let us consider the inverse transform of the product:

$$F(s)G(s) = F(s) \int_0^\infty g(u)e^{-su}\, du$$

$$= \int_0^\infty g(u)F(s)e^{-su}\, du$$

$$= \int_0^\infty g(u)\left[\int_0^\infty e^{-st}u(t-u)f(t-u)\, dt\right] du \qquad (5.1)$$

by Theorem 2.1. At this point it should be emphasized that the interchange of the order of integration cannot always be carried out for improper integrals. It can be shown, however, that if f and g are sectionally continuous and of exponential order, the interchange is valid. As a result, integral (5.1) becomes

$$\int_0^\infty \left[\int_0^\infty u(t-u)f(t-u)g(u)\, du\right] e^{-st}\, dt. \qquad (5.2)$$

111

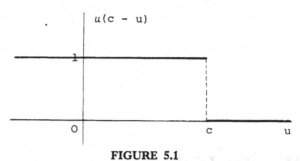

FIGURE 5.1

Clearly, the function $u(c - u) = u[-(u - c)]$ is identically 0 to the right of c (Figure 5.1), so that integral (5.2) reduces to

$$\int_0^\infty \left[\int_0^t f(t - u)g(u) \, du\right] e^{-st} \, dt. \tag{5.3}$$

The last integral suggests the following definition:

Definition 5.1. The *convolution* of two functions f and g is defined by

$$f * g = \int_0^t f(t - u)g(u) \, du. \tag{5.4}$$

For the convolution of two functions we have proved

Theorem 5.1. Convolution Theorem. If $f(t)$ and $g(t)$ are sectionally continuous and of exponential order $e^{\alpha t}$, then

$$L\{f * g\} = F(s)G(s) \qquad (s > \alpha).$$

If we replace $t - u$ by v in equation (5.4), we have

$$f * g = \int_0^t f(v)g(t - v) \, dv = g * f. \tag{5.5}$$

Observe also that in the special case where $g(t) \equiv 1$,

$$L\{f(t) * 1\} = L\left\{\int_0^t f(u)\, du\right\} = \frac{1}{s} F(s). \tag{5.6}$$

Example 5.1

$$t * t^2 = \int_0^t (t - u)u^2\, du$$

$$= \int_0^t (tu^2 - u^3)\, du$$

$$= t\, \frac{u^3}{3} - \frac{u^4}{4}\,\Big|_0^t$$

$$= \frac{t^4}{3} - \frac{t^4}{4}$$

$$= \frac{t^4}{12}.$$

The convolution theorem can be useful in obtaining certain inverse transforms, as illustrated in the following examples.

Example 5.2

$$L^{-1}\left\{\frac{1}{(s^2 + a^2)^2}\right\} = L^{-1}\left\{\frac{1}{s^2 + a^2} \cdot \frac{1}{s^2 + a^2}\right\}$$

$$= \frac{1}{a^2} L^{-1}\left\{\frac{a}{s^2 + a^2} \cdot \frac{a}{s^2 + a^2}\right\}$$

$$= \frac{1}{a^2} (\sin at) * (\sin at)$$

$$= \frac{1}{a^2} \int_0^t \sin au \sin a(t - u)\, du$$

$$= \frac{1}{2a^3} (\sin at - at \cos at).$$

Example 5.3

$$L^{-1}\left\{\frac{1}{s(s-a)}\right\} = L^{-1}\left\{\frac{1}{s} \cdot \frac{1}{s-a}\right\}$$

$$= 1 * e^{at}$$

$$= \int_0^t e^{au} \, du$$

$$= \frac{1}{a}\left(e^{at} - 1\right).$$

Example 5.4. Find

$$L^{-1}\left\{\frac{1}{(s-1)s^{1/2}}\right\}.$$

From equation (1.6),

$$L^{-1}\left\{\frac{1}{s^{1/2}}\right\} = \frac{1}{(\pi t)^{1/2}};$$

hence

$$L^{-1}\left\{\frac{1}{(s-1)s^{1/2}}\right\} = e^t * \frac{1}{(\pi t)^{1/2}}$$

$$= \int_0^t e^{t-u} \frac{1}{(\pi u)^{1/2}} \, du$$

$$= \frac{e^t}{\pi^{1/2}} \int_0^t \frac{e^{-u}}{u^{1/2}} \, du.$$

Letting $x = u^{1/2}$, $2x \, dx = du$, and we obtain

$$\frac{2e^t}{\pi^{1/2}} \int_0^{t^{1/2}} e^{-x^2} \, dx = e^t \operatorname{erf} t^{1/2},$$

where $\operatorname{erf}(t) = (2/\pi^{1/2}) \int_0^t e^{-x^2} \, dx$. This is called the *error function of t*, also known as the probability integral; it will be encountered again in Chapter 7. Because of the first translation theorem, we have also shown that

$$L\{\text{erf } t^{1/2}\} = \frac{1}{s(s+1)^{1/2}}.$$

Example 5.5. Formally,

$$L^{-1}\{F(s)\} = L^{-1}\{F(s) \cdot 1\}$$
$$= f(t) * \delta(t),$$

suggesting that

$$f(t) * \delta(t) = f(t). \tag{5.7}$$

This can also be seen from equation (2.9).

Another area to which the convolution theorem can be applied is that of *integral equations*, equations in which the unknown function appears under the integral sign. We shall consider the important special case

$$x(t) = f(t) + \int_0^t k(t - u)x(u) \, du,$$

which is called an *integral equation of the convolution type*. Such equations can be transformed into algebraic equations by means of the convolution theorem. For obvious reasons an equation such as

$$x''(t) + 2x'(t) + x(t) + \int_0^t e^{t-u}x(u) \, du = 0$$

is called an *integrodifferential equation*, which can be solved by using both the derivative and convolution theorems.

Example 5.6. Let us solve the integral equation

$$x(t) = 1 + \int_0^t \cos(t - u)x(u)\, du.$$

By the convolution theorem,

$$X(s) = \frac{1}{s} + X(s)\left(\frac{s}{s^2 + 1}\right),$$

so that

$$X(s) = \frac{s^2 + 1}{s(s^2 - s + 1)}$$

$$= \frac{1}{s} + \frac{1}{s^2 - s + 1}$$

$$= \frac{1}{s} + \frac{2}{(3)^{1/2}} \frac{(3)^{1/2}/2}{(s - \frac{1}{2})^2 + \frac{3}{4}},$$

whence

$$x(t) = 1 + \frac{2(3)^{1/2}}{3} e^{t/2} \sin \frac{(3)^{1/2}}{2} t.$$

Exercises

5.1. Use the convolution theorem to find

(a) $L^{-1}\left\{\dfrac{a^2}{s(s^2 + a^2)}\right\}$

(b) $L^{-1}\left\{\dfrac{2as}{(s^2 + a^2)^2}\right\}$

(c) $L^{-1}\left\{\dfrac{2as^2}{(s^2 + a^2)^2}\right\}$

5.2. Show that

$$\int_0^t J_0(u) J_0(t - u) \, du = \sin t.$$

(See Exercise 1.7.)

In Exercises 5.3–5.7, solve the given integral equations for $x(t)$.

5.3. $x(t) = a + \displaystyle\int_0^t x(u) \, du$

5.4. $x(t) = t + \displaystyle\int_0^t (t - u)x(u) \, du$

5.5. $x(t) = c \cos t + \displaystyle\int_0^t \sin(t - u)x(u) \, du$

5.6. $\displaystyle\int_0^t x(u)x(t - u) \, du = at$

5.7. $\displaystyle\int_0^t x(u)x(t - u) \, du = a \sin t$ (see Exercise 1.7)

5.8. Solve the following integrodifferential equation for $i(t)$:

$$0.02 \frac{di(t)}{dt} + 16i(t) + 3200 \int_0^t i(u) \, du = 100, \qquad i(0) = 0.$$

5.9. Show that the transform of the solution of the general integral equation of the convolution type,

$$x(t) = f(t) + \int_0^t k(t - u)x(u) \, du,$$

is given by

$$X(s) = \frac{F(s)}{1 - K(s)}.$$

5.10. Show that the Laplace transform of the *sine integral*

$$\mathrm{Si}(t) = \int_0^t \frac{\sin u}{u} \, du$$

is $(1/s)$ Arccot s by Example 1.25.

5.2. Two Special Limits

Theorem 5.2. If the limits exist,

(a) $\lim_{t \to \infty} f(t) = \lim_{s \to 0} sF(s)$,

(b) $\lim_{t \to 0+} f(t) = \lim_{s \to \infty} sF(s)$.

Proof. (a) Proceeding formally, since

$$\int_0^\infty f'(t)e^{-st} \, dt = sF(s) - f(0+) \tag{5.8}$$

and

$$\lim_{s \to 0} \int_0^\infty f'(t)e^{-st} \, dt = \int_0^\infty f'(t) \, dt$$

$$= \lim_{t \to \infty} \int_0^t f'(t) \, dt$$

$$= \lim_{t \to \infty} [f(t) - f(0+)],$$

it follows that

$$\lim_{s \to 0} [sF(s) - f(0+)] = \lim_{t \to \infty} [f(t) - f(0+)]$$

or

$$\lim_{s \to 0} sF(s) = \lim_{t \to \infty} f(t).$$

(b) From equation (5.8),

$$\lim_{s \to \infty} \int_0^\infty f'(t)e^{-st} \, dt = \lim_{s \to \infty} [sF(s) - f(0+)].$$

Under suitable conditions,

$$\lim_{s \to \infty} \int_0^\infty f'(t)e^{-st} \, dt = \int_0^\infty \lim_{s \to \infty} f'(t)e^{-st} \, dt = 0,$$

which implies that

$$\lim_{s \to \infty} [sF(s) - f(0+)] = 0.$$

Since

$$f(0+) = \lim_{t \to 0^+} f(t),$$

we conclude that

$$\lim_{s \to \infty} sF(s) = \lim_{t \to 0^+} f(t).$$

5.3. Applications

Under suitable initial conditions the differential equation of the electrical circuit in Chapter 1 can be written

$$L \frac{di(t)}{dt} + Ri(t) + \frac{1}{C} \int_0^t i(t) \, dt = e(t),$$

an integrodifferential equation. Suppose a voltage $e(t) = E_0 e^{-at}$ is applied to an inductor L and a capacitor C in series. Suppose further that at $t = 0$ there is no current but that C has a charge of Q_0. Then the equation becomes

$$L \frac{di(t)}{dt} + \frac{1}{C} \int_0^t i(t) \, dt + \frac{Q_0}{C} = E_0 e^{-at}; \qquad (5.9)$$

here we think of Q_0 as the initial charge and of $\int_0^t i(t) \, dt$ as the charge acquired in the interval 0 to t. Taking Laplace transforms of both sides, we get

$$\left(Ls + \frac{1}{Cs}\right) I(s) = \frac{E_0}{s + a} - \frac{Q_0}{Cs}$$

and

$$I(s) = \frac{E_0 s/L}{(s+a)[s^2 + (1/LC)]} - \frac{Q_0/LC}{s^2 + (1/LC)}.$$

Hence

$$i(t) = \frac{E_0/L}{[a^2 + (1/LC)]} \left[\frac{1}{(LC)^{1/2}} \sin \frac{t}{(LC)^{1/2}} \right.$$

$$\left. + a \cos \frac{t}{(LC)^{1/2}} - ae^{-at} \right] - \frac{Q_0}{(LC)^{1/2}} \sin \frac{t}{(LC)^{1/2}}.$$

Recalling that $i = dq/dt$, equation (5.9) could be solved by converting it to the differential equation

$$L \frac{d^2 q(t)}{dt^2} + \frac{1}{C} q(t) = E_0 e^{-at}$$

with initial conditions $i(0) = q'(0) = 0$ and $q(0) = Q_0$. Transforming,

$$Ls^2 Q(s) - sLQ_0 + \frac{1}{C} Q(s) = \frac{E_0}{s+a}$$

and

$$Q(s) = \frac{E_0}{(s+a)[Ls^2 + (1/C)]} + \frac{LQ_0 s}{Ls^2 + (1/C)}$$

$$= \frac{E_0/L}{(s+a)[s^2 + (1/LC)]} + \frac{Q_0 s}{s^2 + (1/LC)}.$$

Now, since $dq/dt = i$, $I(s) = sQ(s) - q(0)$, and we have

$$I(s) = \frac{E_0 s/L}{(s+a)[s^2 + (1/LC)]} + \frac{Q_0 s^2}{s^2 + (1/LC)} - Q_0$$

$$= \frac{E_0 s/L}{(s+a)[s^2 + (1/LC)]} - \frac{Q_0/LC}{s^2 + (1/LC)},$$

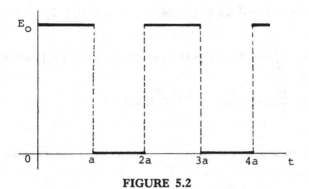

FIGURE 5.2

as before. It is worth noting that the second initial condition is contained in the integrodifferential equation (5.9).

Example 5.7. Solve the equation

$$Ri(t) + \frac{1}{C}\int_0^t i(t)\,dt = e(t), \tag{5.10}$$

where $e(t)$ is the square wave in Figure 5.2.

Since $e(t) = E_0[\varkappa(t) - \varkappa(t - a) + \varkappa(t - 2a) - \cdots]$,

$$RI(s) + \frac{I(s)}{Cs} = E_0\left(\frac{1}{s} - \frac{e^{-as}}{s} + \frac{e^{-2as}}{s} - \cdots\right)$$

and

$$I(s) = \frac{E_0/R}{s + (1/RC)}(1 - e^{-as} + e^{-2as} - \cdots), \tag{5.11}$$

whence

$$i(t) = \frac{E_0}{R}[e^{-t/RC} - \varkappa(t - a)e^{-(t-a)/RC} + \varkappa(t - 2a)e^{-(t-2a)/RC} - \cdots].$$

On the interval $an \leq t < (n + 1)a$, $n = 0, 1, 2, \ldots$,

$$i(t) = \frac{E_0}{R} e^{-t/RC}[1 - e^{a/RC} + e^{2a/RC} - \cdots + (-1)^n e^{na/RC}]$$

$$= \frac{E_0}{R} e^{-t/RC}\left[\frac{1 - (-e^{a/RC})^{n+1}}{1 + e^{a/RC}}\right]. \tag{5.12}$$

As a means of comparison, suppose we write equation (5.10) in the form

$$R\frac{dq(t)}{dt} + \frac{1}{C} q(t) = E_0[u(t) - u(t - a) + u(t - 2a) - \cdots] \tag{5.13}$$

subject to the initial condition $q(0) = 0$. Then

$$RsQ(s) + \frac{1}{C} Q(s) = E_0\left(\frac{1}{s} - \frac{e^{-as}}{s} + \frac{e^{-2as}}{s} - \cdots\right)$$

and

$$Q(s) = \frac{E_0/R}{s[s + (1/RC)]} (1 - e^{-as} + e^{-2as} - \cdots). \tag{5.14}$$

At this point we could convert equation (5.14) into equation (5.11), since $I(s) = sQ(s)$. If we wish to find $q(t)$ instead, we express equation (5.14) as

$$Q(s) = \left[\frac{E_0 C}{s} - \frac{E_0 C}{s + (1/RC)}\right](1 - e^{-as} + e^{-2as} - \cdots).$$

Thus

$$q(t) = E_0 C[u(t) - u(t - a) + u(t - 2a) - \cdots]$$
$$\quad - E_0 C[e^{-t/RC} - u(t - a)e^{-(t-a)/RC} + u(t - 2a)e^{-(t-2a)/RC} - \cdots].$$

Now on the interval $an \leq t < (n+1)a$, $n = 0, 1, 2, \ldots,$

$$i(t) = \frac{dq(t)}{dt} = \frac{E_0}{R} e^{-t/RC}[1 - e^{a/RC} + e^{2a/RC} - \cdots + (-1)^n e^{na/RC}],$$

in agreement with equation (5.12). The first method appears to be simpler since partial fractions have been avoided.

Example 5.8. Consider the mechanical system and corresponding force–voltage analog in Figure 5.3. The mechanical system may be viewed as a schematic representation of an aircraft landing gear, where m_1 is the mass of the body of the plane, m_2 the mass of the wheels, k_2 the flexible tires, k_1 the springs, and c_1 and c_2 the shock absorbers. The respective equations of the two systems are

$$f_1(t) = c_1(v_1 - v_2) + k_1 \int (v_1 - v_2) \, dt + m_1 \frac{dv_1}{dt},$$

$$f_2(t) = c_2 v_2 + c_1(v_2 - v_1) + k_2 \int v_2 \, dt + k_1 \int (v_2 - v_1) \, dt + m_2 \frac{dv_2}{dt}$$

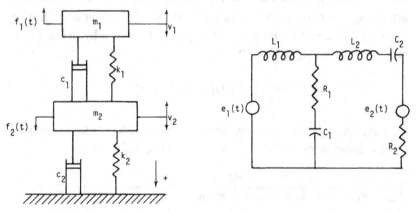

FIGURE 5.3. Schematic diagram of an aircraft landing gear and force–voltage analog.

and

$$e_1(t) = R_1(i_1 - i_2) + \frac{1}{C_1} \int (i_1 - i_2)\, dt + L_1 \frac{di_1}{dt},$$

$$e_2(t) = R_2 i_2 + R_1(i_2 - i_1) + \frac{1}{C_2} \int i_2\, dt + \frac{1}{C_1} \int (i_2 - i_1)\, dt + L_2 \frac{di_2}{dt}.$$

Assuming the initial conditions to be 0, we get the following transformed equations for the mechanical system:

$$F_1(s) = \left(c_1 + \frac{k_1}{s} + m_1 s\right) V_1(s) + \left(-c_1 - \frac{k_1}{s}\right) V_2(s),$$

$$F_2(s) = \left(-c_1 - \frac{k_1}{s}\right) V_1(s) + \left(c_1 + c_2 + \frac{k_1}{s} + \frac{k_2}{s} + m_2 s\right) V_2(s),$$

whence

$$V_1(s) = \frac{F_2(s)(c_1 + k_1/s) + F_1(s)(c_1 + c_2 + k_1/s + k_2/s + m_2 s)}{(c_1 + c_2 + k_1/s + k_2/s + m_2 s)(c_1 + k_1/s + m_1 s) - (c_1 + k_1/s)^2}.$$

Since we are interested in the motion at the instant the plane lands,
let us assume that the effect of the impact is equivalent to imparting
a force $k_2 X_0 u(t)$ on m_2 for some constant X_0. (Think of a car hitting
a curb.) The effect on m_1 at $t = 0$ may then be found by using Theorem 5.2. Assuming $f_1(t) \equiv 0$,

$$\lim_{s \to \infty} s[s^2 V_1(s)] = \lim_{t \to 0^+} v_1''(t).$$

Clearly, $\lim_{t \to 0^+} v_1''(t)$ can be viewed as the sudden jerk experienced
by the passengers at $t = 0$. Now by L'Hospital's rule,

$$\lim_{s \to \infty} \frac{s^3 F_2(s)(c_1 + k_1/s)}{(c_1 + c_2 + k_1/s + k_2/s + m_2 s)(c_1 + k_1/s + m_1 s) - (c_1 + k_1/s)^2}$$

$$= \frac{X_0 c_1 k_2}{m_1 m_2}.$$

According to this result, the smoothness of the landing is improved by flexible tires with small k_2 as well as by small damping c_1 and large masses.

Remark 5.1. This example illustrates again how the solution of a problem can be studied without actually finding the inverse transform. Recall that we determined the motion of the weight on a spring, including the existence of the damping factor, by looking only at the transform. Similarly, the Hurwitz criterion in Chapter 2 provides a method for determining if solutions contain exponential decaying terms from the nature of the zeros in the denominator.

Exercises

5.11. In the discussion of the landing gear, assume that $f_1(t) = -F_0$ (i.e., a constant upward lift). To study the effect of the landing on the body of the plane, assume that $f_2(t) = X_0 k_2 \delta(t)$, and examine $\lim_{t \to 0+} v_1'(t)$.

5.12. Consider an *LRC* circuit with zero initial conditions and $e(t) = E_0 \delta(t)$. Find the current as a function of time, and compare the result with that of Exercise 2.33.

5.13. Consider the response of an *RC* circuit $[q(0) = 0]$ to a single square wave or pulse (see the following diagram).

    (a)   Find the transform of the output voltage from $I(s)$.
    (b)   Find the output voltage $e_2(t)$.

(This is another example of a differentiating circuit.)

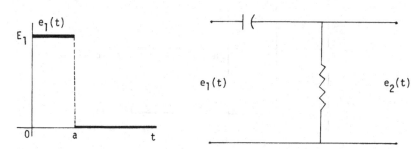

5.14. Assume that the initial charge across the capacitor in the circuit in the following diagram is zero. As in Exercise 5.13 the input is a single square wave voltage pulse of height E_1. Find the transform of the output. [Recall that with zero initial conditions $Q(s) = I(s)/s$; this is an example of an *integrating circuit*.]

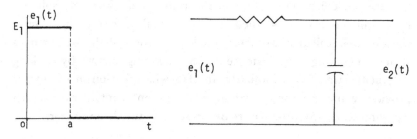

5.15. A unit impulse of voltage is applied to an *RC* circuit initially in a quiescent state. Find the current.

5.16. Find the steady state of the output voltage $Ri(t)$ in Example 5.7.

In Exercises 5.17–5.20 the initial charge across the capacitor in an *RC* circuit is assumed to be zero; find the current in each case.

5.17. $e(t) = [u(t) - u(t - \pi/\omega)] \cos \omega t.$

5.18. $e(t)$ is the square wave in the following diagram.

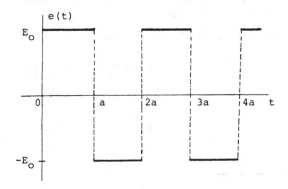

5.19. $e(t)$ is the sawtooth function in the following diagram.

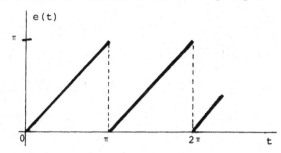

5.20. $e(t) = |\sin \omega t|$.

6

Transforms with Infinitely Many Singularities

6.1. Introduction

In this chapter we are going to study a type of transform not yet encountered. First, let us recall that in Chapter 4 all of the functions had only a finite number of poles. In this chapter we shall consider a more complicated case, one in which $F(s)$ has infinitely many poles; in such a case the denominator of $F(s)$ cannot be a simple polynomial. To find the inverse transform we proceed by constructing the curve in Figure 4.1 with radius R so as to include only finitely many singularities and then taking the limit as $R \to \infty$.

To illustrate the method, suppose we reconsider some problems from earlier sections.

6.2. Examples

Example 6.1. Work Exercise 2.47.
For $v(0) = 0$ we have

$$m \frac{dv}{dt} = P[\delta(t) + \delta(t - a) + \delta(t - 2a) + \cdots].$$

Taking transforms,

$$sV(s) = \frac{P}{m}\,(1 + e^{-as} + e^{-2as} + \cdots)$$

and

$$V(s) = \frac{P}{m}\left(\frac{1}{s} + \frac{e^{-as}}{s} + \frac{e^{-2as}}{s} + \cdots\right). \qquad (6.1)$$

Then

$$v(t) = \frac{P}{m}\,[u(t) + u(t-a) + u(t-2a) + \cdots], \qquad (6.2)$$

the solution obtained earlier.

We now proceed to find $v(t)$ in a different form. From equation (6.1),

$$V(s) = \frac{P}{ms} \cdot \frac{1}{1 - e^{-as}},$$

so that

$$v(t) = \frac{1}{2\pi i}\int_{c-i\infty}^{c+i\infty} \frac{P}{m}\,\frac{e^{st}}{s(1 - e^{-as})}\,ds \qquad (6.3)$$

by the complex inversion formula. To evaluate this integral we must first locate the poles. As in the case of functions with finitely many singularities, we start by finding those values of s for which the denominator of the integrand in equation (6.3) is 0. Clearly,

$$1 - e^{-as} = 0$$

whenever

$$as = 2n\pi i, \qquad n = 0, \pm 1, \pm 2, \ldots,$$

since

$$e^{-2n\pi i} = \cos 2n\pi - i\sin 2n\pi = 1.$$

Consequently, for $n \neq 0$ we get a simple pole at $s = 2n\pi i/a$. If $n = 0$, both factors in the denominator in equation (6.3) are zero, implying that we have a double pole at $s = 0$. This can also be seen by noting that

$$\frac{e^{st}}{s(1 - e^{-as})} = \frac{1 + st + s^2t^2/2! + \cdots}{s(1 - 1 + sa - s^2a^2/2! + \cdots)}$$

$$= \frac{1 + st + s^2t^2/2! + \cdots}{s^2(a - sa^2/2! + \cdots)}$$

$$= \frac{1/a}{s^2} + \text{terms of higher degree.}$$

The residue at $s = 0$ is

$$\frac{1}{1!} \lim_{s \to 0} \frac{d}{ds}\left[\frac{s^2 e^{st}}{s(1 - e^{-as})}\right] = \lim_{s \to 0} \frac{d}{ds}\left(\frac{s e^{st}}{1 - e^{-as}}\right) = \frac{t}{a} + \frac{1}{2}.$$

Next we consider the residue at $s = 2n\pi i/a$ for $n \neq 0$:

$$\lim_{s \to 2n\pi i/a} \frac{(s - 2n\pi i/a)e^{st}}{s(1 - e^{-as})} = \lim_{s \to 2n\pi i/a} \frac{s - 2n\pi i/a}{1 - e^{-as}} \cdot \lim_{s \to 2n\pi i/a} \frac{e^{st}}{s}.$$

Applying L'Hospital's rule, we get

$$\lim_{s \to 2n\pi i/a} \frac{1}{ae^{-as}} \cdot \lim_{s \to 2n\pi i/a} \frac{e^{st}}{s} = \frac{1}{ae^{-2n\pi i}} \cdot \frac{e^{2n\pi ti/a}}{2n\pi i/a}$$

$$= \frac{e^{2n\pi ti/a}}{2n\pi i}, \qquad n = \pm 1, \pm 2, \ldots,$$

since $e^{-2n\pi i} = \cos 2n\pi - i \sin 2n\pi = 1$.

To complete the problem we consider the curve C_k in Figure 6.1, where R_k is chosen so that only a finite number of poles are inside the curve and none on the curve. More specifically, for R_k the poles inside the curve are at $s = 2n\pi i/a$, $-k \leq n \leq k$, where k is a positive

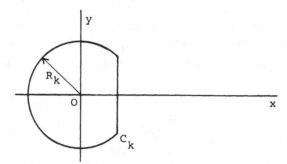

FIGURE 6.1. The basic path of integration containing only finitely many poles.

integer. Thus for C_k,

$$\frac{1}{2\pi i} \int_{C_k} \frac{P}{m} \frac{e^{st}}{s(1 - e^{-as})} \, ds = \frac{P}{m} \left(\frac{t}{a} + \frac{1}{2} + \sum_{\substack{n=-k \\ n \neq 0}}^{k} \frac{e^{2n\pi ti/a}}{2n\pi i} \right).$$

Taking the limit as $k \to \infty$ and recalling that the integral along the circular portion goes to 0, we get

$$L^{-1}\left\{ \frac{P}{ms(1 - e^{-as})} \right\} = \frac{P}{m} \left(\frac{t}{a} + \frac{1}{2} + \sum_{\substack{n=-\infty \\ n \neq 0}}^{\infty} \frac{e^{2n\pi ti/a}}{2n\pi i} \right).$$

The infinite series can be put into a simpler form by eliminating the negative indices as follows:

$$\sum_{\substack{n=-\infty \\ n \neq 0}}^{\infty} = \sum_{n=-\infty}^{-1} + \sum_{n=1}^{\infty}$$

$$= \sum_{n=1}^{\infty} \frac{e^{-2n\pi ti/a}}{-2n\pi i} + \sum_{n=1}^{\infty} \frac{e^{2n\pi ti/a}}{2n\pi i}$$

$$= \sum_{n=1}^{\infty} \frac{1}{\pi n} \frac{e^{2n\pi ti/a} - e^{-2n\pi ti/a}}{2i}$$

$$= \sum_{n=1}^{\infty} \frac{1}{\pi n} \sin \frac{2n\pi t}{a}$$

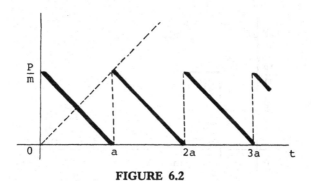

FIGURE 6.2

by equation (3.5). Finally,

$$v(t) = \frac{P}{m}\left(\frac{t}{a} + \frac{1}{2} + \frac{1}{\pi}\sum_{n=1}^{\infty}\frac{1}{n}\sin\frac{2n\pi t}{a}\right). \tag{6.4}$$

To compare this result to equation (6.2) we remark that

$$\frac{P}{m}\left(\frac{1}{2} + \frac{1}{\pi}\sum_{n=1}^{\infty}\frac{1}{n}\sin\frac{2\pi n t}{a}\right)$$

is actually the Fourier series of the periodic function shown in Figure 6.2. When Pt/ma is added to the series, we obtain the step function in Figure 6.3. This step function can be written

$$\frac{P}{m}\left[u(t) + u(t-a) + u(t-2a) + \cdots\right]$$

$$= \frac{nP}{m} \quad \text{for } (n-1)a \le t < na, \; n = 1, 2, 3, \ldots, \tag{6.5}$$

in agreement with equation (6.2). The advantage of form (6.4) is that the solution is a single expression valid for all t (except at the points of discontinuity), while equation (6.5) expresses $v(t)$ only on each interval $(n-1)a \le t < na$.

Example 6.2. Work Exercise 5.20.

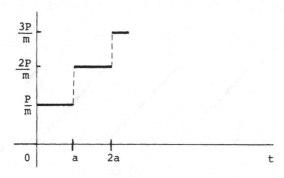

FIGURE 6.3

The equation is

$$Ri(t) + \frac{1}{C} \int_0^t i(t)\, dt = E_0 \, | \sin \omega t \, |.$$

From Exercise 2.46,

$$L\{| \sin \omega t \, |\} = \frac{\omega(1 + e^{-\pi s/\omega})}{(s^2 + \omega^2)(1 - e^{-\pi s/\omega})},$$

so that

$$I(s) = \frac{E_0 C \omega s (1 + e^{-\pi s/\omega})}{(s^2 + \omega^2)(RCs + 1)(1 - e^{-\pi s/\omega})}$$

and

$$i(t) = \frac{E_0 \omega/R}{2\pi i} \int_{c-i\infty}^{c+i\infty} \frac{s(1 + e^{-\pi s/\omega})e^{st}}{(s^2 + \omega^2)(s + a)(1 - e^{-\pi s/\omega})}\, ds, \quad (6.6)$$

where $a = 1/RC$.

As in Example 6.1 we locate the poles by first finding the values of s for which the denominator in equation (6.6) is zero. Now if $s = 2n\omega i$,

$$1 - e^{-2n\pi\omega i/\omega} = 1 - \cos 2n\pi + i \sin 2n\pi$$

$$= 0 \quad \text{for } n = 0, \pm 1, \pm 2, \dots.$$

However, there is no pole at $s = 0$ due to the factor s in the numerator of equation (6.6). To see this, consider

$$\frac{s}{1 - e^{-\pi s/\omega}} = \frac{s}{1 - (1 - \pi s/\omega + \pi^2 s^2/\omega^2 - \cdots)}$$

$$= \frac{1}{\pi/\omega - \pi^2 s/\omega^2 + \cdots}$$

$$= \frac{\omega}{\pi} + \text{other terms},$$

which is a series without negative exponents. Similarly, since $1 + e^{-\pi s/\omega} = 0$ at $s = i\omega$, the factor $s^2 + \omega^2$ does not yield a pole. Consequently, the only poles are at $s = -a$ and $s = 2n\omega i$, $n = \pm 1$, $\pm 2, \ldots$.

For the residue at $s = -a$ we find

$$\frac{E_0\omega}{R} \lim_{s \to -a} \frac{(s + a)(1 + e^{-\pi s/\omega})se^{st}}{(s^2 + \omega^2)(s + a)(1 - e^{-\pi s/\omega})}$$

$$= \frac{E_0\omega}{R} \left[\frac{-ae^{-at}(1 + e^{a\pi/\omega})}{(\omega^2 + a^2)(1 - e^{a\pi/\omega})} \right].$$

At $s = 2n\omega i$, $n = \pm 1, \pm 2, \ldots$,

$$\frac{E_0\omega}{R} \lim_{s \to 2n\omega i} \frac{s - 2n\omega i}{1 - e^{-\pi s/\omega}} \cdot \lim_{s \to 2n\omega i} \frac{s(1 + e^{-\pi s/\omega})e^{st}}{(s^2 + \omega^2)(s + a)}$$

$$= \frac{E_0\omega}{R} \lim_{s \to 2n\omega i} \frac{\omega}{\pi} \frac{1}{e^{-\pi s/\omega}} \frac{s(1 + e^{-\pi s/\omega})e^{st}}{(s^2 + \omega^2)(s + a)}$$

$$= \frac{E_0\omega^2}{R\pi} \frac{e^{2n\omega ti}(4n\omega i)}{\omega^2(1 - 4n^2)(a + 2n\omega i)}.$$

Adding these residues and employing the limiting procedure in Example 6.1, we obtain the series

$$\frac{E_0}{R\pi} \sum_{\substack{n=-\infty \\ n \neq 0}}^{\infty} \frac{e^{2n\omega ti}(4n\omega i)}{(1 - 4n^2)(a + 2n\omega i)}.$$

As before we can simplify this expression by eliminating the negative indices:

$$\frac{E_0}{R\pi}\left[\sum_{n=1}^{\infty}\frac{e^{2n\omega ti}(4n\omega i)}{(1-4n^2)(a+2n\omega i)}-\frac{e^{-2n\omega ti}(4n\omega i)}{(1-4n^2)(a-2n\omega i)}\right]$$

$$=\frac{E_0}{R\pi}\left\{\sum_{n=1}^{\infty}\frac{4n\omega i[e^{2n\omega ti}(a-2n\omega i)-e^{-2n\omega ti}(a+2n\omega i)]}{(1-4n^2)(a^2+4n^2\omega^2)}\right\}.$$

Now by Euler's identity,

$$e^{2n\omega ti}=\cos 2n\omega t+i\sin 2n\omega t$$

and

$$e^{-2n\omega ti}=\cos 2n\omega t-i\sin 2n\omega t,$$

whence

$$\frac{E_0}{R\pi}\sum_{n=1}^{\infty}\frac{4n\omega i(2ai\sin 2n\omega t-4n\omega i\cos 2n\omega t)}{(1-4n^2)(a^2+4n^2\omega^2)}$$

$$=\frac{8E_0\omega}{R\pi}\sum_{n=1}^{\infty}\left(\frac{n}{4n^2-1}\right)\left(\frac{a\sin 2n\omega t-2n\omega\cos 2n\omega t}{a^2+4n^2\omega^2}\right).$$

Finally, combining all the residues,

$$i(t)=\frac{E_0\omega}{R}\left\{\frac{8}{\pi}\sum_{n=1}^{\infty}\left(\frac{n}{4n^2-1}\right)\frac{(1/RC)\sin 2n\omega t-2n\omega\cos 2n\omega t}{(1/R^2C^2)+4n^2\omega^2}\right.$$

$$\left.-\left(\frac{(1/RC)e^{-t/RC}}{\omega^2+(1/R^2C^2)}\right)\left(\frac{1+e^{\pi/RC\omega}}{1-e^{\pi/RC\omega}}\right)\right\}$$

or

$$i(t)=E_0C\omega\left[\frac{8}{\pi}\sum_{n=1}^{\infty}\left(\frac{n}{4n^2-1}\right)\frac{\sin 2n\omega t-2RCn\omega\cos 2n\omega t}{1+4n^2R^2C^2\omega^2}\right.$$

$$\left.-\frac{(1+e^{\pi/RC\omega})e^{-t/RC}}{(1+R^2C^2\omega^2)(1-e^{\pi/RC\omega})}\right].$$

Remark 6.1. To work the remaining examples we need to recall that

$$\sinh i\theta = i \sin \theta$$

and

$$\cosh i\theta = \cos \theta$$

from equations (3.7) and (3.8). These relationships imply that

$$\sinh s = 0 \quad \text{when } s = n\pi i, \qquad n = 0, \pm 1, \pm 2, \ldots,$$

and

$$\cosh s = 0 \quad \text{when } s = (n + \tfrac{1}{2})\pi i, \; n = 0, \pm 1, \pm 2, \ldots.$$

Example 6.3. Do Exercise 2.54; i.e., obtain a different form for the solution of the equation

$$L \frac{di(t)}{dt} + Ri(t) = e(t), \qquad i(0) = 0,$$

where $e(t)$ is the square wave in Figure 6.4.

We know from equation (2.17) that the closed form of the transform of the square wave is $(1/s) \tanh(as/2)$. Hence

$$I(s) = \frac{E_0}{L} \frac{1}{s[s + (R/L)]} \frac{\sinh(as/2)}{\cosh(as/2)}. \tag{6.7}$$

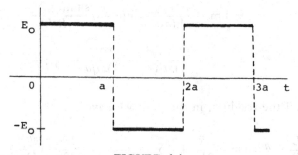

FIGURE 6.4

Since

$$\frac{\sinh(as/2)}{s} = \frac{1}{s}\left(\frac{as}{2} + \frac{a^3 s^3}{3!\,2^3} + \cdots\right)$$

$$= \frac{a}{2} + \frac{a^3 s^2}{3!\,2^3} + \cdots,$$

we see that the factor s in the denominator of equation (6.7) does not lead to a pole at 0. There is, however, a pole at $s = -(R/L)$. Also,

$$\cosh\frac{as}{2} = 0 \quad \text{when} \quad \frac{as}{2} = \left(n + \frac{1}{2}\right)\pi i, \; n = 0, \pm 1, \pm 2, \dots.$$

Omitting E_0/L, the residue at $s = -(R/L)$ is

$$\lim_{s \to -(R/L)} \left(s + \frac{R}{L}\right) \frac{e^{st}\tanh(as/2)}{s(s + R/L)} = \frac{e^{-Rt/L}}{-(R/L)}\tanh\left(-\frac{aR}{2L}\right)$$

$$= \frac{L}{R}\,e^{-Rt/L}\tanh\frac{aR}{2L}.$$

At $s = a_n = (2n + 1)\pi i/a$,

$$\lim_{s \to a_n}(s - a_n)\frac{e^{st}}{s(s + R/L)}\,\frac{\sinh(as/2)}{\cosh(as/2)}$$

$$= \lim_{s \to a_n}\frac{s - a_n}{\cosh(as/2)}\lim_{s \to a_n}\frac{e^{st}\sinh(as/2)}{s(s + R/L)}$$

$$= \frac{2e^{(2n+1)\pi t i/a}}{(2n + 1)\pi i[(2n + 1)\pi i/a + R/L]}.$$

Adding all the residues, in the limit we have

$$i(t) = \frac{L}{R}\,e^{-Rt/L}\tanh\frac{aR}{2L} + \sum_{n=-\infty}^{\infty}\frac{2e^{(2n+1)\pi t i/a}}{(2n + 1)\pi i[(2n + 1)\pi i/a + R/L]}.$$

The series can again be put into a simpler form. First,

$$\sum_{n=-\infty}^{\infty} = \sum_{n=-\infty}^{-1} + \sum_{n=0}^{\infty},$$

so that the first series on the right becomes

$$\sum_{n=1}^{\infty} \frac{2e^{-(2n-1)\pi ti/a}}{-(2n-1)\pi i[R/L - (2n-1)\pi i/a]}.$$

Also, the second series can be written

$$\sum_{n=1}^{\infty} \frac{2e^{(2n-1)\pi ti/a}}{(2n-1)\pi i[R/L + (2n-1)\pi i/a]}.$$

Combining, we get

$$2aL \sum_{n=1}^{\infty} \frac{e^{(2n-1)\pi ti/a}}{(2n-1)\pi i[aR + (2n-1)\pi Li]} - \frac{e^{-(2n-1)\pi ti/a}}{(2n-1)\pi i[aR - (2n-1)\pi Li]}$$

$$= 2aL \sum_{n=1}^{\infty} \frac{e^{(2n-1)\pi ti/a}[aR - (2n-1)\pi Li] - e^{-(2n-1)\pi ti/a}[aR + (2n-1)\pi Li]}{(2n-1)\pi i[a^2R^2 + (2n-1)^2\pi^2 L^2]}.$$

By Euler's identity this reduces to

$$\frac{4aL}{\pi} \sum_{n=1}^{\infty} \frac{aR \sin(2n-1)\pi t/a - (2n-1)\pi L \cos(2n-1)\pi t/a}{(2n-1)[a^2R^2 + (2n-1)^2\pi^2 L^2]}.$$

Finally, multiplying by E_0/L,

$$i(t) = \frac{E_0}{R} e^{-Rt/L} \tanh \frac{aR}{2L}$$

$$+ \frac{4aE_0}{\pi} \sum_{n=1}^{\infty} \frac{aR \sin(2n-1)\pi t/a - (2n-1)\pi L \cos(2n-1)\pi t/a}{(2n-1)[a^2R^2 + (2n-1)^2\pi^2 L^2]}.$$

Example 6.4. Find

$$L^{-1}\left\{\frac{\sinh xs^{1/2}}{s \sinh ls^{1/2}}\right\}.$$

We first observe that

$$\frac{\sinh xs^{1/2}}{s \sinh ls^{1/2}} = \frac{xs^{1/2} + (xs^{1/2})^3/3! + (xs^{1/2})^5/5! + \cdots}{s[ls^{1/2} + (ls^{1/2})^3/3! + (ls^{1/2})^5/5! + \cdots]}$$

$$= \frac{x + sx^3/3! + \cdots}{s[l + sl^3/3! + \cdots]} \tag{6.8}$$

$$= \frac{x/l}{s} + \text{terms of higher order.} \tag{6.9}$$

We can see from equation (6.8) that the function is analytic except where the denominator vanishes. Now

$$\sinh ls^{1/2} = 0 \qquad \text{when } ls^{1/2} = n\pi i, \ n = 0, \pm 1, \pm 2, \ldots,$$

$$s^{1/2} = \frac{n\pi i}{l},$$

and

$$s = -\frac{n^2\pi^2}{l^2} \qquad \text{for } n = 0, 1, 2, \ldots.$$

In addition, the denominator contains the factor s; however, it is clear from equation (6.9) that we have only a simple pole at $s = 0$. For the residue at this point we find

$$\lim_{s\to 0} \frac{se^{st} \sinh xs^{1/2}}{s \sinh ls^{1/2}} = \lim_{s\to 0} \frac{e^{st} \sinh xs^{1/2}}{\sinh ls^{1/2}}$$

$$= \lim_{s\to 0} \frac{\sinh xs^{1/2}}{\sinh ls^{1/2}} \cdot \lim_{s\to 0} e^{st}$$

$$= \lim_{s\to 0} \frac{(x/2s^{1/2}) \cosh xs^{1/2}}{(l/2s^{1/2}) \cosh ls^{1/2}}$$

$$= x/l.$$

At $s = a_n = -(n^2\pi^2/l^2)$, $n = 1, 2, \ldots$,

$$\lim_{s \to a_n} \frac{(s - a_n)e^{st} \sinh xs^{1/2}}{s \sinh ls^{1/2}}$$

$$= \lim_{s \to a_n} \frac{s - a_n}{\sinh ls^{1/2}} \cdot \lim_{s \to a_n} \frac{e^{st} \sinh xs^{1/2}}{s}$$

$$= \lim_{s \to a_n} \frac{1}{l(\cosh ls^{1/2})/2s^{1/2}} \cdot \lim_{s \to a_n} \frac{e^{st} \sinh xs^{1/2}}{s}$$

$$= \frac{1}{[l/(2n\pi i/l)] \cosh(ln\pi i/l)} \cdot \frac{e^{-n^2\pi^2 t/l^2} \sinh(n\pi xi/l)}{-(n^2\pi^2/l^2)}$$

$$= \frac{2n\pi i}{l^2 \cos n\pi} \frac{-l^2 e^{-n^2\pi^2 t/l^2} \sinh(n\pi xi/l)}{n^2\pi^2}$$

$$= \frac{2e^{-n^2\pi^2 t/l^2} \sin(n\pi x/l)}{n\pi \cos n\pi}$$

$$= \frac{2(-1)^n}{n\pi} e^{-n^2\pi^2 t/l^2} \sin \frac{n\pi x}{l}, \qquad n = 1, 2, \ldots.$$

Adding the residues and taking the limit, we obtain

$$L^{-1}\left\{ \frac{\sinh xs^{1/2}}{s \sinh ls^{1/2}} \right\} = \frac{x}{l} + \frac{2}{\pi} \sum_{n=1}^{\infty} \frac{(-1)^n}{n} e^{-n^2\pi^2 t/l^2} \sin \frac{n\pi x}{l}.$$

Example 6.5. Find

$$L^{-1}\left\{ \frac{\cosh xs^{1/2}}{s \cosh s^{1/2}} \right\}.$$

We observe from

$$\frac{\cosh xs^{1/2}}{s \cosh s^{1/2}} = \frac{1 + x^2 s/2! + x^4 s^2/4! + \cdots}{s(1 + s/2! + s^2/4! + \cdots)}$$

that there is a simple pole at $s = 0$. This can be seen from the fact that $\cosh 0 = 1$. Now,

$$\cosh s^{1/2} = 0 \qquad \text{for } s^{1/2} = (n + \tfrac{1}{2})\pi i, \quad n = 0, \pm 1, \pm 2, \ldots,$$

or

$$s = -(n - \tfrac{1}{2})^2 \pi^2, \qquad n = 1, 2, \ldots.$$

The residue at $s = 0$ is

$$\lim_{s \to 0} \frac{s e^{st} \cosh x s^{1/2}}{s \cosh s^{1/2}} = 1,$$

and at $s = a_n = -(n - \tfrac{1}{2})^2 \pi^2$,

$$\lim_{s \to a_n} \frac{s - a_n}{\cosh s^{1/2}} \cdot \lim_{s \to a_n} \frac{e^{st} \cosh x s^{1/2}}{s}$$

$$= \lim_{s \to a_n} \frac{1}{(1/2 s^{1/2}) \sinh s^{1/2}} \cdot \lim_{s \to a_n} \frac{e^{st} \cosh x s^{1/2}}{s}$$

$$= \frac{2(n - \tfrac{1}{2})\pi i}{\sinh(n - \tfrac{1}{2})\pi i} \frac{e^{-(n-1/2)^2 \pi^2 t} \cosh(n - \tfrac{1}{2})\pi i x}{-(n - \tfrac{1}{2})^2 \pi^2}$$

$$= \frac{2(n - \tfrac{1}{2})\pi i}{i \sin(n - \tfrac{1}{2})\pi} \frac{e^{-(n-1/2)^2 \pi^2 t} \cos(n - \tfrac{1}{2})\pi x}{-(n - \tfrac{1}{2})^2 \pi^2}$$

$$= \frac{2}{\pi \sin(n - \tfrac{1}{2})\pi} \frac{e^{-(n-1/2)^2 \pi^2 t} \cos(n - \tfrac{1}{2})\pi x}{-(2n - 1)/2}$$

$$= \frac{4(-1)^n}{(2n - 1)\pi} e^{-(n-1/2)^2 \pi^2 t} \cos\left(n - \frac{1}{2}\right)\pi x.$$

Hence

$$L^{-1}\left\{\frac{\cosh x s^{1/2}}{s \cosh s^{1/2}}\right\} = 1 + \frac{4}{\pi} \sum_{n=1}^{\infty} \frac{(-1)^n}{2n - 1} e^{-(n-1/2)^2 \pi^2 t} \cos\left(n - \frac{1}{2}\right)\pi x.$$

Remark 6.2. An evaluation of the inverse transform of a function with a branch point can be found in Appendix A, Section 4.

Exercises

Most of the following inverse transforms will be needed in the next chapter. Use the complex inversion formula to prove the following relationships:

6.1. $L^{-1}\left\{\dfrac{\sinh xs^{1/2}}{\sinh as^{1/2}}\right\} = \dfrac{2\pi}{a^2}\sum\limits_{n=1}^{\infty}(-1)^n n e^{-n^2\pi^2 t/a^2}\sin\dfrac{n\pi x}{a}$

6.2. $L^{-1}\left\{\dfrac{\sinh sx}{s^2\sinh sa}\right\} = \dfrac{xt}{a} + \dfrac{2a}{\pi^2}\sum\limits_{n=1}^{\infty}\dfrac{(-1)^n}{n^2}\sin\dfrac{n\pi x}{a}\sin\dfrac{n\pi t}{a}$

6.3. $L^{-1}\left\{\dfrac{(-K + K\cosh s^{1/2})\sinh xs^{1/2}}{s\sinh s^{1/2}}\right\}$

$$= \dfrac{4K}{\pi}\sum\limits_{n=1}^{\infty}\dfrac{1}{2n-1}e^{-(2n-1)^2\pi^2 t}\sin(2n-1)\pi x$$

6.4. $L^{-1}\left\{\dfrac{(2 - K + K\cosh s^{1/2})\sinh xs^{1/2}}{s\sinh s^{1/2}}\right\} = 2x +$

$$\dfrac{2}{\pi}\sum\limits_{n=1}^{\infty}\dfrac{(-1)^n}{n}[2 - K + (-1)^n K]e^{-n^2\pi^2 t}\sin n\pi x$$

6.5. $L^{-1}\left\{\dfrac{\cosh x(s/a)^{1/2}}{s\cosh l(s/a)^{1/2}}\right\} = 1 + \dfrac{4}{\pi}\sum\limits_{n=1}^{\infty}\dfrac{(-1)^n}{2n-1}e^{-(2n-1)\pi^2 at/4l^2}\cos\dfrac{(2n-1)\pi x}{2l}$

6.6. $L^{-1}\left\{\dfrac{\pi\sinh(sx/a)}{l(s^2 + \pi^2/l^2)\sinh(sl/a)}\right\} = \dfrac{\sin(\pi x/al)\sin(\pi t/l)}{\sin(\pi/a)}$

$$+ \dfrac{2a}{\pi}\sum\limits_{n=1}^{\infty}\dfrac{(-1)^n}{a^2 n^2 - 1}\sin\dfrac{n\pi x}{l}\sin\dfrac{n\pi at}{l}$$

6.7. $L^{-1}\left\{\dfrac{1 - \cosh(sx/a)}{s^3} + \dfrac{\sinh(sl/a)\sinh(sx/a)}{s^3\cosh(sl/a)}\right\} = \dfrac{x(2l - x)}{2a^2}$

$$- \dfrac{16l^2}{\pi^3 a^2}\sum\limits_{n=1}^{\infty}\dfrac{1}{(2n-1)^3}\sin\dfrac{(2n-1)\pi x}{2l}\cos\dfrac{(2n-1)\pi at}{2l}$$

6.8. $L^{-1}\left\{\dfrac{\sinh xs^{1/2}}{s^{3/2}\cosh s^{1/2}}\right\} = x - \dfrac{8}{\pi^2}\sum\limits_{n=1}^{\infty}\dfrac{(-1)^{n-1}}{(2n-1)^2}e^{-(2n-1)\pi^2 t/4}\sin\dfrac{(2n-1)\pi x}{2}$

7

Applications to Partial Differential Equations

7.1. Introduction

In our earlier work on the spring we dealt with either a single equation or a system of two equations depending on whether one or two point masses were involved. Systems containing an infinite number of discrete point masses are subject to an infinite system of ordinary differential equations. We shall now consider a system with a continuous mass distribution, which leads to a partial differential equation, with independent variables in time and space.

Suppose a perfectly flexible string of length L is stretched along the x axis (Figure 7.1). Let T be the constant tension at any point on the string and ϱ the mass of the string per unit length. We shall make the following assumptions:

(a) The constant tension T is so large compared to the weight of the string that the gravitational force may be neglected.

(b) The displacement $Y(x, t)$ is so small with respect to L that the length of the string may be taken as L at each of its positions.

(c) Each point P on the string traverses a straight line parallel to the Y axis, i.e., the vibrations are purely transverse.

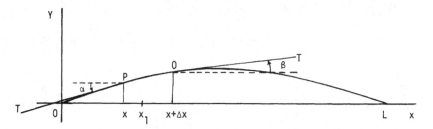

FIGURE 7.1. A perfectly flexible string of length L stretched along the x axis.

These assumptions, though quite severe, are adequately satisfied in the case of, say, strings of musical instruments.

Returning to Figure 7.1, we see that $T(\sin \beta - \sin \alpha)$ is an unbalanced force in the Y direction acting on the segment PQ. If $x < x_1 \leq x + \Delta x$, we have, by Newton's second law,

$$(\varrho \, \Delta x) \frac{\partial^2 Y(x_1, t)}{\partial t^2} = T(\sin \beta - \sin \alpha).$$

For small θ, $\sin \theta \simeq \tan \theta$, so that

$$(\varrho \, \Delta x) \frac{\partial^2 Y(x_1, t)}{\partial t^2} = T(\tan \beta - \tan \alpha)$$

or

$$(\varrho \, \Delta x) \frac{\partial^2 Y(x_1, t)}{\partial t^2} = T\left[\frac{\partial Y(x + \Delta x, t)}{\partial x} - \frac{\partial T(x, t)}{\partial x} \right].$$

Dividing by $\varrho \, \Delta x$ and letting $\Delta x \to 0$, we have (since $x_1 \to x$)

$$\frac{\partial^2 Y(x, t)}{\partial t^2} = \frac{T}{\varrho} \frac{\partial^2 Y(x, t)}{\partial x^2}.$$

A more convenient form of the equation is

$$\frac{\partial^2 Y(x, t)}{\partial t^2} = a^2 \frac{\partial^2 Y(x, t)}{\partial x^2}, \tag{7.1}$$

where $a = (T/\varrho)^{1/2}$, called the *propagation constant*.

The equation can also be obtained as the limit of a sequence of ordinary differential equations. Suppose a set of beads of mass $m = c\varrho$ are strung at equal intervals c on a light string. Let Y_i be the transverse displacement of the ith bead and θ_i the small angle between the string segment joining the ith and $(i + 1)$st beads and the x axis. Now, a derivative can be approximated by a difference quotient and a second derivative by a second difference quotient. (This is a useful procedure in numerical analysis.) Also, since $\sin \theta_i \simeq \tan \theta_i = (Y_{i+1} - Y_i)/c$, the upward force component on the ith bead becomes

$$T \sin \theta_i \simeq T(Y_{i+1} - Y_i)/c$$

from the right and

$$-T \sin \theta_{i-1} \simeq T(Y_{i-1} - Y_i)/c$$

from the left. Hence

$$c\varrho \frac{\partial^2 Y_i}{\partial t^2} = T(Y_{i+1} - 2Y_i + Y_{i-1})/c.$$

If we now divide by $c\varrho$, we obtain a second difference quotient that tends to the second derivative with respect to x if $c \to 0$. At the same time the number of beads becomes large and the mass of each correspondingly small. Thus

$$\frac{\partial^2 Y}{\partial t^2} = \frac{T}{\varrho} \frac{\partial^2 Y}{\partial x^2},$$

as before.

Equation (7.1) is a special case of the general homogeneous equation of second order

$$a \frac{\partial^2 U}{\partial x^2} + 2b \frac{\partial^2 U}{\partial x\, \partial y} + c \frac{\partial^2 U}{\partial y^2} + 2d \frac{\partial U}{\partial x} + 2e \frac{\partial U}{\partial y} + fU = 0, \tag{7.2}$$

where the coefficients may be functions of x and y. The form of equation (7.2) resembles that of a conic section and has led to a similar classification: The equations are called *elliptic*, *hyperbolic*, or *parabolic* according to whether $b^2 - ac$ is less than, greater than, or equal to zero, respectively. Equation (7.1) is an example of a hyperbolic equation.

Since we are going to use Laplace transforms to solve partial differential equations, let us first note that the functions under consideration have two independent variables. So we shall consider the transform with respect to one variable only, thinking of the other as a parameter. For $F(x, t)$, for example, we define the Laplace transform to be

$$\int_0^\infty e^{-st}F(x, t) \, dt \qquad \text{or} \qquad \int_0^\infty e^{-sx}F(x, t) \, dx.$$

In the succeeding sections t is always the time variable, and we shall find it convenient to take the transform of $F(x, t)$ with respect to t. To distinguish this case from the use of the transform in Chapter 1, let us employ the notation

$$L_t\{F(x, t)\} = \int_0^\infty e^{-st}F(x, t) \, dt = f(x, s).$$

Consequently, all properties of the transform derived earlier carry over with only slight modifications in the notation. For example, the derivative theorem now reads

$$L_t\left\{\frac{\partial}{\partial t} F(x, t)\right\} = sL_t\{F(x, t)\} - F(x, 0+). \tag{7.3}$$

However, in the case of $(\partial/\partial x)F(x, t)$ we observe that

$$L_t\left\{\frac{\partial}{\partial x} F(x, t)\right\} = \int_0^\infty e^{-st} \frac{\partial}{\partial x} F(x, t) \, dt$$

$$= \frac{\partial}{\partial x} \int_0^\infty e^{-st}F(x, s) \, dt,$$

which implies that

$$L_t\left\{\frac{\partial}{\partial x} F(x,\, t)\right\} = \frac{\partial}{\partial x} f(x,\, s) \tag{7.4}$$

whenever the order of integration and differentiation can be interchanged.

Example 7.1

$$L_t\left\{\frac{\partial}{\partial t} \sin xt\right\} = sL_t\{\sin xt\} - 0$$

$$= \frac{xs}{s^2 + x^2}.$$

Example 7.2. Suppose we check equation (7.4) for $F(x,\, t) = \sin xt$. First we evaluate $L_t\{(\partial/\partial x) \sin xt\}$ directly:

$$L_t\left\{\frac{\partial}{\partial x} \sin xt\right\} = L_t\{t \cos xt\} = -\frac{\partial}{\partial s}\left(\frac{s}{s^2 + x^2}\right)$$

by Theorem 1.6. Hence

$$L_t\left\{\frac{\partial}{\partial x} \sin xt\right\} = \frac{s^2 - x^2}{(s^2 + x^2)^2}. \tag{7.5}$$

Now by equation (7.4),

$$L_t\left\{\frac{\partial}{\partial x} \sin xt\right\} = \frac{\partial}{\partial x} f(x,\, s)$$

$$= \frac{\partial}{\partial x}\left(\frac{x}{s^2 + x^2}\right)$$

$$= \frac{s^2 - x^2}{(s^2 + x^2)^2},$$

in agreement with equation (7.5).

To see how Laplace transforms can be used to solve partial differential equations, let us consider the following artificial problem:

$$\frac{\partial U(x, t)}{\partial x} + 2\frac{\partial U(x, t)}{\partial t} = 0 \qquad (x > 0, \ t > 0), \qquad (7.6)$$

$$\text{(a)} \quad U(0+, t) = 0 \qquad (t \geq 0),$$

$$\text{(b)} \quad U(x, 0) = 1 \qquad (x > 0).$$

By equations (7.3) and (7.4) the transformed equation becomes

$$\frac{\partial u(x, s)}{\partial x} + 2su(x, s) - 2U(x, 0) = 0.$$

This is an ordinary differential equation if $u(x, s)$ is regarded as a function of x alone, s being a parameter. (For convenience we shall use the symbol for ordinary differentiation.) By condition (b),

$$\frac{d}{dx} u(x, s) + 2su(x, s) = 2;$$

the general solution of this equation is

$$u(x, s) = ce^{-2sx} + 1/s,$$

where c depends on the parameter s. By condition (a),

$$c = -\frac{1}{s},$$

so that

$$u(x, s) = -\frac{1}{s} e^{-2xs} + \frac{1}{s}.$$

Taking inverse transforms,

$$U(x, t) = 1 - u(t - 2x)$$

or

$$U(x, t) = \begin{cases} 1, & t < 2x \\ 0, & t \geq 2x. \end{cases} \tag{7.7}$$

Looking at the solution of the equation, some observations are in order. First, the solution is not continuous along the line $t = 2x$ in the xt plane. Not being differentiable there, function (7.7) does not satisfy the equation along this line, but only on each side of the line. If we now let $x \to 0^+$,

$$U(0+, t) = 1 - u(t) = 0,$$

so that condition (a) is satisfied. For $t = 0$,

$$U(x, 0) = 1 - u(-2x),$$

and since $x > 0$, $u(-2x) = 0$, and condition (b) is also satisfied. Yet because of the discontinuity, we have to confess that we have not found the solution to the equation. In physical problems, such as the vibrating string, one is, nevertheless, left with the feeling that a solution ought to exist. So if the function $U(x, t)$ satisfies the equation and the auxiliary conditions except along certain curves, $U(x, t)$ is called a *generalized solution*, otherwise a *strict solution*. We shall check each case separately without attempting to discuss the general situation. Another observation is that uniqueness of the solution cannot be asserted, since the function

$$U(x, t) = \begin{cases} 1, & t < x \\ 0, & t \geq x \end{cases}$$

satisfies the equation and the conditions equally well, except along the line $t = x$. Apparently the conditions in this problem are not restrictive enough to yield a unique solution.

Clearly, then, the auxiliary conditions play a key role in partial differential equations. Since t is ordinarily a time variable, condition (b) in equation (7.6) is called an *initial condition*. This type of condition will naturally be imposed when the differential equation governs the motion of a system. Condition (a) involves the space variable x and is called a *boundary condition*. Boundary conditions prescribe the physical behavior of a system at the frontier or boundary of the region.

We can see from our artificial example that the auxiliary conditions must be

(1) weak enough so that a solution exists and
(2) strong enough to ensure uniqueness.

Finally, to solve partial differential equations by Laplace transforms, (1) the equation must be linear, (2) the coefficients must be constant, and (3) at least one of the variables must have a range from 0 to ∞. In solving equations, we shall concentrate mostly on the actual method of solution without attempting to prove the uniqueness of the solutions obtained.

7.2. The Diffusion Equation

As another illustration of the transform method, consider the following partial differential equation of parabolic type:

$$\frac{\partial U(x, t)}{\partial t} = k \frac{\partial^2 U(x, t)}{\partial x^2} \qquad (x > 0,\ t > 0), \qquad (7.8)$$

subject to the initial condition

$$\text{(a)} \quad U(x, 0+) = U_0 \qquad (x > 0),$$

and the boundary conditions

$$\text{(b)} \quad U(0, t) = U_1 \qquad (t > 0),$$

$$\text{(c)} \quad \lim_{x \to \infty} U(x, t) = U_0 \qquad (t > 0).$$

By equations (7.3) and (7.4) the transformed equation becomes

$$su(x, s) - U(x, 0+) = k \frac{\partial^2}{\partial x^2} u(x, s), \tag{7.9}$$

which is an ordinary differential equation if $u(x, s)$ is regarded as a function of x alone. Employing the symbol for ordinary differentiation again and making use of initial condition (a) in equation (7.9), we obtain

$$\frac{d^2}{dx^2} u(x, s) - \frac{s}{k} u(x, s) = -\frac{U_0}{k} \tag{7.10}$$

whose solution is

$$u(x, s) = c_1 e^{(s/k)^{1/2}x} + c_2 e^{-(s/k)^{1/2}x} + \frac{U_0}{s} \tag{7.11}$$

(c_1 and c_2 depend on the parameter s). Since by boundary condition (c) $\lim_{x \to \infty} u(x, s) = U_0/s$, we must have $c_1 = 0$. Also, condition (b) implies that $u(0, s) = U_1/s$; this can be substituted into equation (7.11) to yield

$$c_2 = \frac{U_1}{s} - \frac{U_0}{s}.$$

Consequently, the solution of equation (7.10) is given by

$$u(x, s) = \frac{U_1 - U_0}{s} e^{-(s/k)^{1/2}x} + \frac{U_0}{s}. \tag{7.12}$$

The transform (7.12) looks unfamiliar. It is, however, related to the error function from Chapter 5. Recalling that

$$\text{erf}(t) = \frac{2}{\pi^{1/2}} \int_0^t e^{-x^2} \, dx,$$

we now define

$$\text{erfc}(t) = 1 - \text{erf}(t),$$

called the *complementary error function*. It is shown in Appendix A (Section 4), that

$$L\left\{\text{erfc}\left(\frac{a}{2t^{1/2}}\right)\right\} = \frac{e^{-as^{1/2}}}{s}. \tag{7.13}$$

Applying formula (7.13) to (7.12) (with the help of Theorem 1.2), we obtain the solution of the boundary-value problem:

$$U(x, t) = U_0 + (U_1 - U_0)\,\text{erfc}\,\frac{x}{2(kt)^{1/2}}. \tag{7.14}$$

To check if the function (7.14) really satisfies the equation, let us write it in the form

$$U(x, t) = U_0 + (U_1 - U_0)\left(1 - \frac{2}{\pi^{1/2}}\int_0^{x/2(kt)^{1/2}} e^{-u^2}\,du\right)$$

and find the necessary derivatives. [Note that $U(x, t)$ is continuous on $(0, \infty)$ together with its first- and second-order partial derivatives.]

$$\frac{\partial U(x, t)}{\partial t} = (U_1 - U_0)\left(-\frac{2}{\pi^{1/2}}\right)e^{-x^2/4kt}\frac{\partial}{\partial t}\left[\frac{x}{2(kt)^{1/2}}\right]$$

$$= \frac{(U_1 - U_0)xe^{-x^2/4kt}}{2t(\pi kt)^{1/2}},$$

$$\frac{\partial U(x, t)}{\partial x} = (U_1 - U_0)\left(-\frac{2}{\pi^{1/2}}\right)e^{-x^2/4kt}\frac{\partial}{\partial x}\left[\frac{x}{2(kt)^{1/2}}\right]$$

$$= -\frac{(U_1 - U_0)e^{-x^2/4kt}}{(\pi kt)^{1/2}},$$

and

$$\frac{\partial^2 U(x, t)}{\partial x^2} = -\frac{U_1 - U_0}{(\pi kt)^{1/2}}e^{-x^2/4kt}\left(-\frac{x}{2kt}\right) = \frac{(U_1 - U_0)xe^{-x^2/4kt}}{2kt(\pi kt)^{1/2}},$$

or

$$\frac{\partial^2 U(x, t)}{\partial x^2} = \frac{1}{k}\frac{\partial U(x, t)}{\partial t}.$$

Since $(2/\pi^{1/2})\int_0^\infty e^{-u^2}\,du = 1$ (Section 1.3), it is easy to check that conditions (a) and (c) are satisfied, while $U(0, t) = U_1$, in agreement with condition (b). This shows that function (7.14) satisfies the equation and all conditions.

The partial differential equation

$$\frac{\partial U}{\partial t} = k\frac{\partial^2 U}{\partial x^2},$$

sometimes called the *diffusion equation*, arises in a number of physical problems. If $U(x, t)$ is the temperature of a solid, then the solution of the equation represents the heat flow as a function of both distance and time. In this case k is called the *thermal diffusivity* of the material and is equal to $a/c\varrho$. Here a is the thermal conductivity, c the specific heat per unit mass, and ϱ the density of the material.[*]

If $U(x, t)$ is the concentration of dissolved matter in a solvent, then the solution stands for the diffusion as a function of distance and time. Now k is called the *diffusion coefficient*.

Finally, the equation may be viewed as a special case of the equation for the flow of electricity in a transmission line. This time $k = 1/RC$ and $U(x, t)$ may stand either for V or I, the potential or current, respectively. Thus

$$\frac{\partial^2 V}{\partial x^2} = RC\frac{\partial V}{\partial t}$$

and

$$\frac{\partial^2 I}{\partial x^2} = RC\frac{\partial I}{\partial t},$$

known as the *telegraph equations*.

---

[*] For a derivation of the heat equation, see R. V. Churchill, *Fourier Series and Boundary Value Problems*, McGraw-Hill, London (1963), pp. 10–13.

Remark 7.1. Before continuing, we need to derive two more special transforms. Since

$$L^{-1}\left\{\frac{e^{-xs^{1/2}}}{s}\right\} = \text{erfc}\,\frac{x}{2t^{1/2}}$$

by equation (7.13) and since $\text{erfc}(x/2t^{1/2}) = 0$ for $t = 0$,

$$L^{-1}\{e^{-xs^{1/2}}\} = L^{-1}\left\{s\,\frac{e^{-xs^{1/2}}}{s}\right\}$$

$$= \frac{d}{dt}\,\text{erfc}\,\frac{x}{2t^{1/2}}$$

$$= \frac{x}{2(\pi t^3)^{1/2}}\,e^{-x^2/4t}$$

by the derivative theorem. We have shown that

$$L^{-1}\{e^{-xs^{1/2}}\} = \frac{x}{2(\pi t^3)^{1/2}}\,e^{-x^2/4t}. \tag{7.15}$$

From this result and Theorem 1.6,

$$L^{-1}\left\{\frac{d}{ds}\,e^{-xs^{1/2}}\right\} = L^{-1}\left\{\frac{-x}{2s^{1/2}}\,e^{-xs^{1/2}}\right\}$$

$$= \frac{-tx}{2(\pi t^3)^{1/2}}\,e^{-x^2/4t},$$

which implies that

$$L^{-1}\left\{\frac{e^{-xs^{1/2}}}{s^{1/2}}\right\} = \frac{1}{(\pi t)^{1/2}}\,e^{-x^2/4t}. \tag{7.16}$$

Example 7.3. Let us find the temperature distribution in the semi-infinite solid $x \geq 0$ (Figure 7.2), whose initial temperature is zero. Since $U(x, t)$ denotes the temperature, the initial condition is $U(x, 0+) = 0$. With this physical model it is natural to assume that

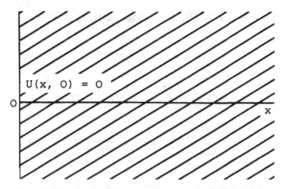

FIGURE 7.2. A semi-infinite solid.

$\lim_{x\to\infty} U(x, t) = 0$. Assume further that $U(0, t) = F(t)$; that is, the face of the solid has temperature $F(t)$ starting at $t = 0$. Letting $k = 1$, we now have the following boundary-value problem:

$$\frac{\partial U(x, t)}{\partial t} = \frac{\partial^2 U(x, t)}{\partial x^2} \qquad (x > 0, \ t > 0),$$

(a) $\quad U(x, 0+) = 0 \qquad\qquad (x > 0),$

(b) $\quad U(0, t) = F(t) \qquad\qquad (t \geq 0),$

(c) $\quad \lim_{x\to\infty} U(x, t) = 0 \qquad (t > 0).$

Taking Laplace transforms and making use of condition (a), we have

$$\frac{d^2}{dx^2} u(x, s) = su(x, s),$$

$$\frac{d^2}{dx^2} u(x, s) - su(x, s) = 0.$$

The solution of the last equation is

$$u(x, s) = c_1 e^{xs^{1/2}} + c_2 e^{-xs^{1/2}}.$$

From condition (c), $c_1 = 0$, and since $u(0, s) = f(s)$,

$$u(x, s) = f(s)e^{-xs^{1/2}}.$$

By equation (7.15) and the convolution theorem,

$$U(x, t) = \int_0^t F(t - u) \frac{x}{2(\pi u^3)^{1/2}} e^{-x^2/4u} \, du. \tag{7.17}$$

It is easily seen that function (7.17) satisfies the equation and conditions (a) and (c). To check the remaining condition, consider

$$H(x, t) = \frac{x}{2(\pi t^3)^{1/2}} e^{-x^2/4t}$$

in the integrand of equation (7.17). If $t > 0$, $H(0, t) = 0$. If $t = 0$, $H(0, t)$ does not exist. To show this, consider

$$\lim_{\substack{x \to 0 \\ t \to 0}} H(x, t)$$

along the path $t = x$. On this path

$$H(x, t) = \frac{1}{2(\pi t)^{1/2}e^{t/4}},$$

which goes to infinity as $t \to 0$. We see, then, that $H(0, t) = \delta(t)$, and, as a result,

$$U(0, t) = F(t) * \delta(t) = F(t)$$

by equation (5.7); thus condition (b) is satisfied.

7.3. The Vibrating String

Suppose a string is stretched along the x axis with the left end at the origin, and suppose $Y(x, t)$ is the transverse displacement. Then $Y(x, t)$ satisfies the partial differential equation

$$\frac{\partial^2 Y}{\partial t^2} = a^2 \frac{\partial^2 Y}{\partial x^2} \qquad (x > 0, \ t > 0),$$

where $a = (T/\varrho)^{1/2}$. (See Section 7.1.) T is the tension on the string and ϱ the mass per unit length.

Example 7.4. If a semi-infinite string is stretched along the x axis and if the initial displacement and velocity are zero, then a natural boundary condition is $\lim_{x \to \infty} Y(x, t) = 0$. Suppose that the end of the string is given an oscillatory motion equal to $\sin t$ starting at $t = 0$. Then the boundary-value problem is

$$\frac{\partial^2 Y(x, t)}{\partial t^2} = a^2 \frac{\partial^2 Y(x, t)}{\partial x^2} \qquad (x > 0, \ t > 0),$$

(a) $\quad Y(x, 0) = 0$ $\qquad\qquad\qquad$ $(x > 0)$,

(b) $\quad \partial Y(x, 0)/\partial t = 0$ $\qquad\qquad$ $(x > 0)$,

(c) $\quad Y(0, t) = \sin t$ $\qquad\qquad\quad$ $(t \geq 0)$,

(d) $\quad \lim_{x \to \infty} Y(x, t) = 0$ $\qquad\quad$ $(t \geq 0)$.

Transforming and making use of the initial conditions (a) and (b), we have

$$s^2 y(x, s) - s Y(x, 0) - \frac{\partial Y(x, 0)}{\partial t} = a^2 \frac{\partial^2 y(x, s)}{\partial x^2}$$

and

$$\frac{d^2}{dx^2} y(x, s) - \frac{s^2}{a^2} y(x, s) = 0.$$

The solution of the resulting differential equation is

$$y(x, s) = c_1 e^{(s/a)x} + c_2 e^{-(s/a)x}.$$

As before, boundary condition (d) implies that $c_1 = 0$, so that

$$y(x, s) = c_2 e^{-(s/a)x}.$$

From condition (c), $y(0, s) = 1/(s^2 + 1)$, whence

$$y(x, s) = \frac{1}{s^2 + 1} e^{-(x/a)s}$$

and

$$Y(x, t) = u\left(t - \frac{x}{a}\right) \sin\left(t - \frac{x}{a}\right).$$

Equivalently,

$$Y(x, t) = \begin{cases} \sin(t - x/a), & t \geq x/a \\ 0, & t < x/a. \end{cases}$$

Since $\sin t$ is continuous and $\sin(0) = 0$, it follows that $Y(x, t)$ is continuous when $x \geq 0$ and $t \geq 0$. Taking second derivatives, the equation is easily seen to be satisfied except along the line $t - x/a = 0$, where the first partial derivatives have a finite jump discontinuity. Also, $Y(0, t) = \sin t$; now, since $u(t - x/a) = 0$ for $t - x/a < 0$ and since this is true for $x > at$, condition (d) is satisfied. Finally, since $Y(x, t) = 0$ for $t < x/a$, $Y(x, 0) = 0$ for $x > 0$. For the same reason $\partial Y(x, 0)/\partial t = 0$.

Example 7.5. Suppose a string of length l is fixed at both ends and has an initial displacement of zero. Assume that the string is given an initial velocity equal to $\sin(\pi x/l)$. Then the equation and conditions are

$$\frac{\partial^2 Y(x, t)}{\partial t^2} = a^2 \frac{\partial^2 Y(x, t)}{\partial x^2} \qquad (0 < x < l,\ t > 0),$$

(a) $\quad Y(x, 0) = 0$ $\qquad\qquad\qquad (x > 0)$,

(b) $\quad \partial Y(x, 0)/\partial t = \sin(\pi x/l) \quad (0 \leq x \leq l)$,

(c) $\quad Y(0, t) = 0$ $\qquad\qquad\qquad (t > 0)$,

(d) $\quad Y(l, t) = 0$ $\qquad\qquad\qquad (t > 0)$.

The transformed equation is

$$s^2 y(x, s) - sY(x, 0) - \frac{\partial Y(x, 0)}{\partial t} = a^2 \frac{\partial^2}{\partial x^2} y(x, s)$$

or

$$\frac{d^2}{dx^2} y(x, s) - \frac{s^2}{a^2} y(x, s) = -\frac{1}{a^2} \sin \frac{\pi x}{l}.$$

Solving,

$$y(x, s) = c_1 \cosh \frac{s}{a} x + c_2 \sinh \frac{s}{a} x + \frac{1}{a^2} \left(\frac{1}{\pi^2/l^2 + s^2/a^2} \right) \sin \frac{\pi x}{l}.$$

By condition (c), $y(0, s) = 0$, implying that $c_1 = 0$, and by condition (d),

$$0 = y(l, s) = c_2 \sinh \frac{s}{a} l,$$

which means that c_2 is also 0. Hence

$$y(x, s) = \frac{1}{s^2 + (a^2\pi^2/l^2)} \sin \frac{\pi x}{l}$$

and

$$Y(x, t) = \frac{l}{a\pi} \sin \frac{a\pi t}{l} \sin \frac{\pi x}{l}.$$

Exercises

7.1. Suppose a semi-infinite solid is bounded by $x = 0$ and that its initial temperature is U_1, a constant. Suppose further that at $t = 0$ the face $x = 0$ is suddenly brought to a constant temperature $U_0 > U_1$ and held at this temperature for $t > 0$. Find the temperature at any point for $t > 0$ if $\lim_{x \to \infty} U(x, t) = U_1$.

7.2. Repeat Exercise 7.1 for the following corresponding conditions:

    (1) $U(x, 0+) = 0$,

    (2) $U(0, t) = U_0$,

    (3) $\lim_{x \to \infty} U(x, t) = 0$.

7.3. Solve the boundary-value problem

$$\frac{\partial U}{\partial t} = k \frac{\partial^2 U}{\partial x^2} \qquad (x > 0,\ t > 0),$$

subject to the conditions

    (1) $U(1, t) = Q_0 \delta(t)$,

    (2) $U(x, 0+) = 0$,

    (3) $\lim_{x \to \infty} U(x, t) = 0$.

[Use equation (7.15) and Theorem 1.2.] If the solid is a semi-infinite bar whose lateral surface is insulated, interpret the problem physically.

7.4. Suppose that the face of a semi-infinite solid has a heat flux $F(t)$ applied at the face $x = 0$; that is, $\partial U(0, t)/\partial x = -F(t)$. If the initial temperature is 0 and $\lim_{x \to \infty} U(x, t) = 0$, find the temperature at any point for all $t > 0$. Assume $k = 1$ and use equation (7.16).

7.5. Suppose $U(x, t)$ is the amount of dissolved matter in a solvent. Solve the following boundary-value problems, and give a physical interpretation:

    (a) $\dfrac{\partial U}{\partial t} = \dfrac{\partial^2 U}{\partial x^2} \qquad (x > 0,\ t > 0)$,

        (1) $U(x, 0+) = 0$ $(x > 0)$,

        (2) $\lim_{x \to \infty} U(x, t) = 0$ $(t > 0)$,

        (3) $\dfrac{\partial U(0, t)}{\partial x} = -K$ $(t \geq 0)$.

(b) $\dfrac{\partial U}{\partial t} = \dfrac{\partial^2 U}{\partial x^2}$ $(x > 0,\ t > 0),$

(1) $U(x, 0+) = 0$ $(x > 0),$

(2) $\lim_{x \to \infty} U(x, t) = 0$ $(t > 0),$

(3) $\dfrac{\partial U(0, t)}{\partial x} = -K\delta(t)$ $(t \geq 0).$

[Check the solutions, especially condition (3).]

7.6. Suppose $V(x, t)$ is the voltage in a semi-infinite telegraph line, and assume that the voltage is 0 initially and that a constant voltage V_0 is applied to the end (at $x = 0$). Find $V(x, t)$; the equation is

$$\frac{\partial^2 V(x, t)}{\partial x^2} = RC\,\frac{\partial V(x, t)}{\partial t} \qquad (x > 0,\ t > 0).$$

7.7. A semi-infinite string is stretched along the x axis, i.e., $Y(x, 0) = \partial Y(x, 0)/\partial t = 0$ for $x > 0$. Suppose the left end $x = 0$ is given a displacement described by the function $G(t)$ with $G(0) = 0$ and that the motion dies out along the string. Find $Y(x, t)$.

7.8. Suppose a string of length l is stretched between two points and held fixed at the ends. Suppose further that the initial displacement is given by $Y(x, 0) = \sin(\pi x/l)$ $(0 \leq x \leq l)$. Find the subsequent motion if $\partial Y(x, 0)/\partial t = 0$.

7.9. Suppose that a string is initially at rest, $\lim_{x \to \infty} Y(x, t) = 0$, and that the left end of the string is given a displacement

$$Y(0, t) = [\alpha(t) - \alpha(t - 1)]t.$$

Find the motion of the string.

7.10. Repeat Exercise 7.9 for $Y(0, t) = \delta(t)$.

7.11. Suppose that a semi-infinite bar whose lateral surface is insulated is bounded by $x = 0$ and that its initial temperature is 0. Suppose further that the left face $x = 0$ is brought to a constant temperature K at $t = 0$ and back down to 0 at $t = 1$. Find the temperature distribution in the bar.

7.4. The Complex Inversion Formula Again

In this section we are going to apply the techniques from Chapter 6 to the solution of boundary-value problems.

Example 7.6. Suppose a bar of length l has an insulated lateral surface and an initial temperature of 0 (Figure 7.3). Suppose further that the temperature at the left end is held at 0 but that the right end has a constant temperature of F_0 for $t > 0$. Letting k be unity, the boundary-value problem can be written

$$\frac{\partial U(x, t)}{\partial t} = \frac{\partial^2 U(x, t)}{\partial x^2} \qquad (0 < x < l, \; t > 0),$$

(a) $U(x, 0) = 0,$

(b) $U(0, t) = 0,$

(c) $U(l, t) = F_0.$

Note: An equally good model for this problem is a large flat slab of thickness l.

Taking transforms, we get

$$su(x, s) = \frac{d^2 u(x, s)}{dx^2}.$$

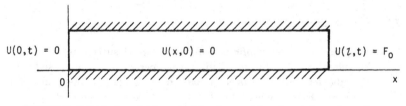

FIGURE 7.3. A bar of length l having an insulated lateral surface.

The solution is

$$u(x, s) = c_1 \sinh x s^{1/2} + c_2 \cosh x s^{1/2}.$$

Since $u(0, s) = 0$, $c_2 = 0$. From the condition $u(l, s) = F_0/s$,

$$c_1 = \frac{F_0}{s \sinh l s^{1/2}}$$

and

$$u(x, s) = \frac{F_0 \sinh x s^{1/2}}{s \sinh l s^{1/2}}.$$

By Example 6.4,

$$U(x, t) = F_0 \left[\frac{x}{l} + \frac{2}{\pi} \sum_{n=1}^{\infty} \frac{(-1)^n}{n} e^{-n^2 \pi^2 t/l^2} \sin \frac{n\pi x}{l} \right],$$

which is the temperature distribution in the bar.

Since

$$\frac{\partial}{\partial t} e^{-n^2 \pi^2 t/l^2} = -\frac{n^2 \pi^2}{l^2} e^{-n^2 \pi^2 t/l^2}$$

and

$$\frac{\partial^2}{\partial x^2} \sin \frac{n\pi x}{l} = -\frac{n^2 \pi^2}{l^2} \sin \frac{n\pi x}{l},$$

the equation is (formally) satisfied, while $U(0, t) = 0$ and $U(l, t) = F_0$ by direct substitution. To check the remaining condition we note that the Fourier series

$$\frac{2}{\pi} \sum_{n=1}^{\infty} \frac{(-1)^n}{n} \sin \frac{n\pi x}{l}$$

converges to $-(x/l)$ on the interval $0 \leq x < l$, so that $U(x, 0) = 0$ for $0 \leq x < l$.

Exercises

7.12. A bar of length l with an insulated lateral surface has an initial temperature of $2x$, while the ends are kept fixed at zero. Find the temperature distribution in the bar, using Example 6.4. Letting k be unity, the boundary-value problem is

$$\frac{\partial U}{\partial t} = \frac{\partial^2 U}{\partial x^2} \qquad (0 < x < l,\ t > 0),$$

(1) $U(x, 0) = 2x,$

(2) $U(0, t) = 0,$

(3) $U(l, t) = 0.$

7.13. Repeat Exercise 7.12 for a bar of length 1 and initial temperature K. (Use Exercise 6.3.)

7.14. In this problem we have a bar (lateral surface insulated) of length 1 and initial temperature K. While the left end is again kept at 0, the right end is held at temperature 2 for $t > 0$. Letting $k = 1$, find the temperature at any point x at any time t. (Use Exercise 6.4.)

7.15. A slab $0 \le x \le l$ has an initial temperature U_0, while the right end has a fixed temperature U_1. The left end is insulated, a fact that can be described by the boundary condition $\partial U(0, t)/\partial x = 0$. (We used a similar condition in Exercises 7.4 and 7.5.) Find the temperature distribution using Exercise 6.5. The boundary-value problem can be written

$$\frac{\partial U}{\partial t} = k\,\frac{\partial^2 U}{\partial x^2} \qquad (0 < x < l,\ t > 0),$$

(1) $U(x, 0) = U_0,$

(2) $\partial U(0, t)/\partial x = 0,$

(3) $U(l, t) = U_1.$

7.16. A string of length l is initially at rest. Suppose the left end is held fixed, while the right end is given a transverse motion described by the function $\sin(\pi t/l)$. Find the motion of the string; the boundary-value problem can be stated as

$$\frac{\partial^2 Y}{\partial t^2} = a^2 \frac{\partial^2 Y}{\partial x^2} \qquad (0 \le x < l,\ t > 0),$$

(1) $Y(x, 0) = Y_t(x, 0) = 0,$

(2) $Y(0, t) = 0,$

(3) $Y(l, t) = \sin(\pi t/l).$

(Use Exercise 6.6.)

7.17. If $Y(x, t)$ is the motion of a vertical bar of length l held fixed at the top and vibrating under its own weight, then $Y(x, t)$ satisfies the equation

$$\frac{\partial^2 Y}{\partial t^2} = a^2 \frac{\partial^2 Y}{\partial x^2} + g \qquad (0 < x < l,\ t > 0)$$

with $\partial Y(l, t)/\partial x = 0$. Use Exercise 6.7 to solve the boundary-value problem if the other conditions are $Y(0, t) = 0$ and $Y(x, 0) = Y_t(x, 0) = 0$.

7.18. Using Exercise 6.8, solve the boundary-value problem

$$\frac{\partial U}{\partial t} = \frac{\partial^2 U}{\partial x^2} \qquad (0 < x < 1,\ t > 0),$$

(1) $U(x, 0) = 0,$

(2) $U(0, t) = 0,$

(3) $\partial U(1, t)/\partial x = -b/K.$

Interpret the problem physically, and compare it to Exercises 7.4 and 7.5.

7.19. Solve the following boundary-value problem, and give a heat-flow interpretation:

$$\frac{\partial U}{\partial t} = \frac{\partial^2 U}{\partial x^2} \qquad (0 < x < 1,\ t > 0),$$

(1) $U(x, 0) = 0,$

(2) $U(0, t) = 0,$

(3) $U(1, t) = Kt.$

7.20. A bar of length l has its end $x = 0$ fixed with all parts of the bar initially at rest and unstrained. A constant force per unit area acts parallel to the bar at the other end. Find the longitudinal displacement $V(x, t)$. The boundary-value problem is

$$\frac{\partial^2 V(x, t)}{\partial t^2} = a^2 \frac{\partial^2 V(x, t)}{\partial x^2} \qquad (0 < x < l,\ t > 0,\ a^2 = E/\varrho),$$

(1) $V(x, 0) = \partial V(x, 0)/\partial t = 0$,

(2) $V(0, t) = 0$,

(3) $\partial V(l, t)/\partial x = F_0/E$ (E is Young's modulus).

7.21. Determine the motion of the free end $x = l$ of the bar in Exercise 7.20. (*Hint*: Use Example 2.13.)

Appendix A

More on Complex Variable Theory

A.1. If $w = f(z)$ is single-valued, the derivative of $f(z)$ is defined by

$$f'(z) = \lim_{\Delta z \to 0} \frac{f(z + \Delta z) - f(z)}{\Delta z}$$

if the limit exists and is independent of the way in which $\Delta z \to 0$. Suppose we write

$$f(z) = u(x, y) + iv(x, y).$$

Then

$$f'(z) = \lim_{\substack{\Delta x \to 0 \\ \Delta y \to 0}} \frac{u(x + \Delta x, y + \Delta y) - u(x, y) + i[v(x + \Delta x, y + \Delta y) - v(x, y)]}{\Delta x + i\,\Delta y}.$$

Next, we let $\Delta x + i\,\Delta y \to 0$ along two different routes. If $\Delta y = 0$, we get

$$\lim_{\Delta x \to 0} \left[\frac{u(x + \Delta x, y) - u(x, y)}{\Delta x} + i\,\frac{v(x + \Delta x, y) - v(x, y)}{\Delta x} \right]$$

$$= \frac{\partial u}{\partial x} + i\,\frac{\partial v}{\partial x}.$$

169

Similarly, for $\Delta x = 0$, we obtain

$$-i\,\frac{\partial u}{\partial y} + \frac{\partial v}{\partial y}.$$

Since the two limits must be the same, it follows that

$$\frac{\partial u}{\partial x} = \frac{\partial v}{\partial y} \quad \text{and} \quad \frac{\partial v}{\partial x} = -\frac{\partial u}{\partial y}.$$

The equations are called the *Cauchy–Riemann equations* and provide a necessary condition that $f(z)$ be analytic in some region. (It can be shown that the condition is also sufficient if the partial derivatives are continuous.)

A.2. If we choose the opposite direction for the path of integration, the integral (3.9) becomes

$$\int_1^0 (t + it)^2(1 + i)\,dt = -(-\tfrac{2}{3} + \tfrac{2}{3}i).$$

It is obvious from this example that choosing the opposite direction for a path of integration changes the sign of the integral.

From this property one can deduce that the value of an integral of an analytic function is independent of the path chosen, given *Cauchy's theorem*:

Theorem (Cauchy–Goursat). Let $f(z)$ be analytic in a region D. Let C be a simple closed curve of finite length in D such that $f(z)$ is analytic inside and on C. Then

$$\int_C f(z)\,dz = 0.$$

(For a proof, see any standard text on complex variable theory.)

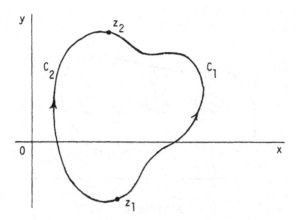

FIGURE A.1. Two different paths of integration between z_1 and z_2.

Consider the integral of $f(z)$ from z_1 to z_2 along C_1 and C_2, respectively (Figure A.1). Denote by $-C_2$ the path from z_2 to z_1. Then

$$\int_{C_1} f(z)\, dz + \int_{-C_2} f(z)\, dz = 0$$

by Cauchy's theorem. Therefore

$$\int_{C_1} f(z)\, dz - \int_{C_2} f(z)\, dz = 0$$

and

$$\int_{C_1} f(z)\, dz = \int_{C_2} f(z)\, dz.$$

A.3. To show that for the shaded region in Figure 3.5

$$\int_C f(z)\, dz = \int_{C_1} f(z)\, dz + \int_{C_2} f(z)\, dz + \int_{C_3} f(z)\, dz,$$

consider the integral along the continuous path shown in Figure A.2.

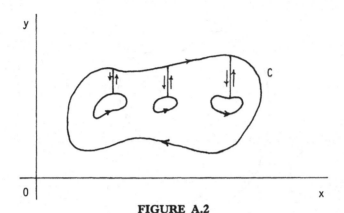

FIGURE A.2

On the region inside this path $f(z)$ is analytic, so by Cauchy's theorem (Section 2) the integral around this path is 0. We observe also that the integrals along the straight lines add to 0 since each integral is taken twice, but in opposite directions. Consequently,

$$\int_C f(z)\,dz + \int_{C_1} f(z)\,dz + \int_{C_2} f(z)\,dz + \int_{C_3} f(z)\,dz = 0$$

or

$$\int_C f(z)\,dz = -\int_{C_1} f(z)\,dz - \int_{C_2} f(z)\,dz - \int_{C_3} f(z)\,dz.$$

Furthermore, the orientations of the paths inside C are opposite of those in Figure 3.5. If we reverse the directions of the integrals along the curves inside, we reverse the signs of these integrals, and the conclusion follows.

A.4. So far all of our functions have been single-valued; i.e., if $w = f(z)$, then to every z there corresponds only one w. Now consider $w = z^{1/2}$, letting $z = re^{i\theta}$, $w = r^{1/2}e^{i\theta/2}$. If θ changes to $\theta + 2\pi$, w changes from $r^{1/2}e^{i\theta/2}$ to $r^{1/2}e^{i(\theta+2\pi)/2} = -r^{1/2}e^{i\theta/2}$. But $z = re^{i\theta}$ and $z = re^{i(\theta+2\pi)}$ represent the same complex number, so that $w = z^{1/2}$ is clearly not single-valued. Now, making another complete circuit about the origin, we get $w = r^{1/2}e^{i(\theta+4\pi)/2} = r^{1/2}e^{i\theta/2}$, which does correspond to the original value. Consequently, we can make $w = z^{1/2}$ single-valued by restricting θ to an interval such as $0 \leq \theta < 2\pi$ (or $-\pi \leq \theta < \pi$). We obtain another single-valued function on the interval $2\pi \leq \theta < 4\pi$. Each choice of θ defines one *branch* of the function $w = z^{1/2}$. If we draw a ray OL from the origin, we can keep the function single-valued by agreeing not to cross this ray. OL is called a *branch line* and the origin a *branch point*.

Suppose we employ these ideas to find

$$f(t) = L^{-1}\left\{\frac{e^{-as^{1/2}}}{s}\right\} = \frac{1}{2\pi i}\int_{c-i\infty}^{c+i\infty}\frac{e^{st-as^{1/2}}}{s}\,ds.$$

By the above discussion there is a branch point at $s = 0$. Hence by

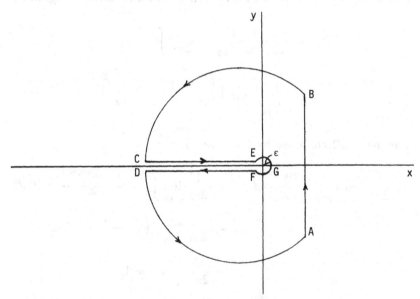

FIGURE A.3. A path of integration which avoids the branch point at $s = 0$ and the branch line along the negative real axis.

restricting the value of θ we can define one branch of the function. For convenience we take $-\pi \leq \theta < \pi$, although other choices are possible, and choose the path in Figure A.3.

Let us note that in the terminology introduced above the negative real axis is a branch line. The parallel portions of the path are actually the same line but are drawn separately for easier visualization. Also, we take the argument of s to be π on CE and $-\pi$ on FD; this way we have avoided crossing the branch line. We have also avoided the singularity by drawing a small circle of radius ε around $s = 0$. Since there are no other singularities, the integral around the entire curve is zero. And since

$$\left| \frac{e^{-as^{1/2}}}{s} \right| < |s|^{-1} = |Re^{i\theta}|^{-1} = \frac{1}{R}$$

on the curved portions BC and DA, the integral along these curves approaches 0 as $R \to \infty$ (by Section 4.2). Hence (omitting integrands)

$$f(t) = \lim_{\substack{R \to \infty \\ \varepsilon \to 0}} \frac{1}{2\pi i} \int_{AB} = \frac{1}{2\pi i} \int_{c-i\infty}^{c+i\infty}$$

$$= -\lim_{\substack{R \to \infty \\ \varepsilon \to 0}} \frac{1}{2\pi i} \left(\int_{CE} + \int_{EGF} + \int_{FD} \right).$$

For the small circle we have $s = \varepsilon e^{i\theta}$ and

$$-\frac{1}{2\pi i} \lim_{\varepsilon \to 0} \int_{EGF} \frac{e^{st - as^{1/2}}}{s} \, ds = -\frac{1}{2\pi i} \lim_{\varepsilon \to 0} \int_{\pi}^{-\pi} \frac{e^{\varepsilon t e^{i\theta} - a\varepsilon^{1/2} e^{i\theta/2}}}{\varepsilon e^{i\theta}} \, i\varepsilon e^{i\theta} \, d\theta$$

$$= -\frac{1}{2\pi i} \lim_{\varepsilon \to 0} i \int_{\pi}^{-\pi} e^{\varepsilon t e^{i\theta} - a\varepsilon^{1/2} e^{i\theta/2}} \, d\theta$$

$$= -\frac{1}{2\pi} \int_{\pi}^{-\pi} d\theta = 1.$$

Along CE and FD we put $s = xe^{i\pi}$ and $s = xe^{-i\pi}$, respectively. In the first case $s^{1/2} = x^{1/2}e^{i\pi/2} = ix^{1/2}$ and

$$\int_{CE} \frac{e^{st-as^{1/2}}}{s} \, ds = \int_{-R}^{-\varepsilon} \frac{e^{st-as^{1/2}}}{s} \, ds = \int_{R}^{\varepsilon} \frac{e^{-xt-aix^{1/2}}}{x} \, dx,$$

since x goes from R to ε as s goes from $-\varepsilon$ to $-R$. Similarly,

$$\int_{FD} \frac{e^{st-as^{1/2}}}{s} \, ds = \int_{\varepsilon}^{R} \frac{e^{-xt+aix^{1/2}}}{x} \, dx.$$

Adding the integrals, we get

$$\int_{\varepsilon}^{R} \frac{e^{-xt}(e^{aix^{1/2}} - e^{-aix^{1/2}})}{x} \, dx = 2i \int_{\varepsilon}^{R} \frac{e^{-xt}\sin ax^{1/2}}{x} \, dx$$

and

$$-\frac{1}{2\pi i} \lim_{\substack{R \to \infty \\ \varepsilon \to 0}} 2i \int_{\varepsilon}^{R} \frac{e^{-xt}\sin ax^{1/2}}{x} \, dx = -\frac{1}{\pi} \int_{0}^{\infty} \frac{e^{-xt}\sin ax^{1/2}}{x} \, dx.$$

Since it is known that

$$\frac{1}{\pi} \int_{0}^{\infty} \frac{e^{-xt}\sin ax^{1/2}}{x} \, dx = \mathrm{erf}\left(\frac{a}{2t^{1/2}}\right),$$

we shall simply state the final result:

$$L^{-1}\left\{\frac{e^{-as^{1/2}}}{s}\right\} = 1 - \mathrm{erf}\left(\frac{a}{2t^{1/2}}\right) = \mathrm{erfc}\left(\frac{a}{2t^{1/2}}\right),$$

where

$$\mathrm{erf}(t) = \frac{2}{\pi^{1/2}} \int_{0}^{t} e^{-u^2} \, du,$$

the error function encountered earlier.

A.5. Show that

$$\int_C \frac{dz}{(z-a)^n} = \begin{cases} 0, & n \neq 1 \\ 2\pi i, & n = 1. \end{cases}$$

[See equation (3.11).] If C is a simple closed curve about $z = a$, then by Section 3,

$$\int_C \frac{dz}{(z-a)^n} = \int_R \frac{dz}{(z-a)^n},$$

where R is a circle centered at a with radius ε small enough so that the circle lies entirely inside C. We now proceed by direct calculation. Since the equation of the circle is

$$z = a + \varepsilon e^{i\theta} \qquad (0 \leq \theta < 2\pi),$$

we get

$$dz = \varepsilon i e^{i\theta}\, d\theta,$$

and

$$\int_R \frac{dz}{(z-a)^n} = \int_0^{2\pi} \frac{i\varepsilon e^{i\theta}}{\varepsilon^n e^{n\theta i}}\, d\theta$$

$$= \frac{i}{\varepsilon^{n-1}} \int_0^{2\pi} e^{(1-n)\theta i}\, d\theta$$

$$= \frac{i}{\varepsilon^{n-1}} \left. \frac{e^{(1-n)\theta i}}{(1-n)i} \right|_0^{2\pi}$$

$$= \frac{i}{\varepsilon^{n-1}(1-n)i} \left[\cos(1-n)\theta + i\sin(1-n)\theta \right]_0^{2\pi}$$

$$= \frac{1}{(1-n)\varepsilon^{n-1}} \left[\cos(1-n)2\pi - 1 \right]$$

$$= 0 \qquad \text{for } n \neq 1.$$

If $n = 1$, the integral reduces to

$$i \int_0^{2\pi} d\theta = 2\pi i.$$

A.6. In this section we shall derive the formula for calculating the residue a_{-1} (Chapter 3). If $f(z)$ has a simple pole at $z = a$, then

$$f(z) = \frac{a_{-1}}{z - a} + a_0 + a_1(z - a) + a_2(z - a)^2 + \cdots.$$

Hence

$$a_{-1} = \lim_{z \to a} (z - a) f(z).$$

If $f(z)$ has a pole of order n at $z = a$, then

$$g(z) = (z - a)^n f(z)$$

is analytic in some neighborhood of $z = a$. By Taylor's formula,

$$g(z) = g(a) + g(a)(z - a) + \cdots + \frac{g^{(n-1)}(a)}{(n-1)!} (z - a)^{n-1} + \cdots.$$

Dividing by $(z - a)^n$, we get directly

$$a_{-1} = \frac{g^{(n-1)}(a)}{(n-1)!}$$

or

$$a_{-1} = \lim_{z \to a} \frac{1}{(n-1)!} \frac{d^{n-1}}{dz^{n-1}} [(z - a)^n f(z)].$$

A.7. (Refer to Chapter 4.) Let $f(z)$ be analytic in a region D. Let C be a simple closed curve of finite length in D such that $f(z)$ is analytic inside and on C. Then the value of $f(z)$ is determined at a point z_0 inside C by the values of $f(z)$ on C:

$$f(z_0) = \frac{1}{2\pi i} \int_C \frac{f(z)}{z - z_0}\, dz.$$

This is known as *Cauchy's integral formula*. For a function analytic in the half-plane $\text{Re}(z) \geq c$, the formula becomes

$$f(z_0) = \frac{-1}{2\pi i} \int_{c-i\infty}^{c+i\infty} \frac{f(z)}{z - z_0}\, dz \qquad [\text{Re}(z) > c],$$

provided that the integral exists. Next, suppose $f(s)$, the Laplace transform of some $F(t)$, is analytic in the half-plane $\text{Re}(s) \geq c$ Proceeding formally, since

$$f(s) = \frac{1}{2\pi i} \int_{c-i\infty}^{c+i\infty} \frac{f(z)}{s - z}\, dz,$$

$$L^{-1}\{f(s)\} = L^{-1}\left\{\frac{1}{2\pi i} \int_{c-i\infty}^{c+i\infty} \frac{f(z)}{s - z}\, dz\right\},$$

and

$$F(t) = \frac{1}{2\pi i} \int_{c-i\infty}^{c+i\infty} e^{zt} f(z)\, dz.$$

Appendix B

Table of Laplace Transforms

$f(t)$	$F(s)$
$e^{at} f(t)$	$F(s - a)$
$f(at)$	$\dfrac{1}{a} F\left(\dfrac{s}{a}\right)$
$f'(t)$	$sF(s) - f(0+)$
$f^{(n)}(t)$	$s^n F(s) - s^{n-1} f(0+) - s^{n-2} f'(0+)$ $- \cdots - f^{(n-1)}(0+)$
$f * g = \displaystyle\int_0^t f(t - u)g(u)\,du$	$F(s)G(s)$
$\displaystyle\int_0^t f(u)\,du$	$\dfrac{1}{s} F(s)$
$t^n f(t),\quad n$ a positive integer	$(-1)^n \dfrac{d^n}{ds^n} F(s)$
$\dfrac{f(t)}{t}$	$\displaystyle\int_s^\infty F(u)\,du$

$f(t)$	$F(s)$
$u(t-a)f(t-a)$	$e^{-as}F(s)$
$f(t+T)=f(t)$	$\dfrac{1}{1-e^{-sT}}\displaystyle\int_0^T e^{-st}f(t)\,dt$
1	$\dfrac{1}{s}$
t	$\dfrac{1}{s^2}$
t^α	$\dfrac{\Gamma(\alpha+1)}{s^{\alpha+1}}$ $(\alpha>-1)$
$t^{-(1/2)}$	$\left(\dfrac{\pi}{s}\right)^{1/2}$
e^{at}	$\dfrac{1}{s-a}$
$\sin at$	$\dfrac{a}{s^2+a^2}$
$\cos at$	$\dfrac{s}{s^2+a^2}$
$e^{bt}\sin at$	$\dfrac{a}{(s-b)^2+a^2}$
$e^{bt}\cos at$	$\dfrac{s-b}{(s-b)^2+a^2}$
$e^{at}t^n$	$\dfrac{n!}{(s-a)^{n+1}}$
$\sinh at$	$\dfrac{a}{s^2-a^2}$

$f(t)$	$F(s)$
$\cosh at$	$\dfrac{s}{s^2 - a^2}$
$at - \sin at$	$\dfrac{a^3}{s^2(s^2 + a^2)}$
$1 - \cos at$	$\dfrac{a^2}{s(s^2 + a^2)}$
$\sin at - at \cos at$	$\dfrac{2a^3}{(s^2 + a^2)^2}$
$\sin at + at \cos at$	$\dfrac{2as^2}{(s^2 + a^2)^2}$
$t \sin at$	$\dfrac{2as}{(s^2 + a^2)^2}$
$t \cos at$	$\dfrac{s^2 - a^2}{(s^2 + a^2)^2}$
$\sinh at + \sin at$	$\dfrac{2as^2}{s^4 - a^4}$
$\sinh at - \sin at$	$\dfrac{2a^3}{s^4 - a^4}$
$\cosh at + \cos at$	$\dfrac{2s^3}{s^4 - a^4}$
$\cosh at - \cos at$	$\dfrac{2a^2s}{s^4 - a^4}$
$J_0(at)$	$\dfrac{1}{(s^2 + a^2)^{1/2}}$
$u(t - t_0)$	$\dfrac{e^{-st_0}}{s}$

$f(t)$	$F(s)$
$\delta(t - t_0)$	e^{-st_0}
$\delta(t)$	1
$\mathrm{Si}(t)$	$\dfrac{1}{s} \,\mathrm{Arccot}\, s$
$\dfrac{1 - e^{-t}}{t}$	$\ln\left(1 + \dfrac{1}{s}\right)$
$\mathrm{erf}\, t^{1/2}$	$\dfrac{1}{s(s + 1)^{1/2}}$
$\mathrm{erfc}\, \dfrac{a}{2t^{1/2}}$	$\dfrac{e^{-as^{1/2}}}{s}$
$\dfrac{a}{2(\pi t^3)^{1/2}}\, e^{-a^2/4t}$	$e^{-as^{1/2}}$
$\dfrac{e^{-a^2/4t}}{(\pi t)^{1/2}}$	$\dfrac{e^{-as^{1/2}}}{s^{1/2}}$

Square wave

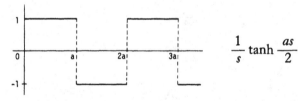

$$\frac{1}{s} \tanh \frac{as}{2}$$

Triangular wave

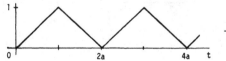

$$\frac{1}{as^2} \tanh \frac{as}{2}$$

$f(t)$	$F(s)$

Full-wave rectification of sine wave
$f(t) = |\sin \omega t|$

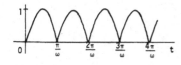

$$\frac{\omega}{s^2 + \omega^2} \coth \frac{\pi s}{2\omega}$$

Half-wave rectification of sine wave

$$\frac{\omega}{s^2 + \omega^2}\left(\frac{1}{1 - e^{-\pi s/\omega}}\right)$$

Sawtooth function

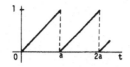

$$\frac{1}{as^2} - \frac{e^{-as}}{s(1 - e^{-as})}$$

$$\frac{x}{l} + \frac{2}{\pi}\sum_{n=1}^{\infty}\frac{(-1)^n}{n}e^{-n^2\pi^2 t/l^2}\sin\frac{n\pi x}{l}$$

$$\frac{\sinh xs^{1/2}}{s \sinh ls^{1/2}}$$

$$1 + \frac{4}{\pi}\sum_{n=1}^{\infty}\frac{(-1)^n}{2n-1}e^{-(n-1/2)^2\pi^2 t}$$
$$\times \cos\left(n - \frac{1}{2}\right)\pi x$$

$$\frac{\cosh xs^{1/2}}{s \cosh s^{1/2}}$$

Other infinite-series forms are listed in the exercises at the end of Section 6.2.

Bibliography

CARSLAW, H. S., *Operational Methods in Applied Mathematics*, Oxford University Press, New York (1942).

CHURCHILL, R. V., *Modern Operational Mathematics in Engineering*, McGraw-Hill, New York (1944).

CHURCHILL, R. V., *Fourier Series and Boundary Value Problems*, McGraw-Hill, London (1963).

DOETSCH, G., *Theorie und Anwendung der Laplace-Transformation*, Springer, Berlin (1937).

GOLDMAN, S., *Laplace Transform Theory and Electrical Transients*, Dover, New York (1966).

HOLBROOK, J. G., *Laplace Transforms for Electronic Engineers*, Pergamon Press, London (1966).

KAPLAN, W., *Advanced Calculus*, Addison-Wesley, Reading, Massachusetts (1959).

MIKUSINSKI, J., *Operational Calculus*, Pergamon Press, Elmsford, N.Y. (1959).

RAINVILLE, E. D., *The Laplace Transform: An Introduction*, Macmillan, New York (1963).

RAINVILLE, E. D., and BEDIENT, P. E., *Elementary Differential Equations*, Macmillan, New York (1974).

SMITH, M. G., *Laplace Transform Theory*, Van Nostrand Reinhold, London (1966).

Answers to Exercises

Section 1.3

1.7. $L\{J_0(t)\} = \dfrac{1}{(s^2 + 1)^{1/2}}$

Section 1.5

1.8. (a) $\dfrac{1}{(s-2)^2}$; (b) $\dfrac{3}{(s+4)^2}$

1.10. (a) $\dfrac{a}{(s-b)^2 + a^2}$; (b) $\dfrac{s-b}{(s-b)^2 - a^2}$

1.11. (a) $e^{2t} \sin 3t$; (b) $2e^{2t} \sin 3t$

1.12. $e^{-3t}(\cos 2t - \frac{3}{2}\sin 2t)$; $e^{-3t}(\cos 4t - \frac{1}{2}\sin 4t)$

1.13. $e^{t/2}\left(\cos \dfrac{(11)^{1/2}}{2}t - \dfrac{1}{(11)^{1/2}}\sin \dfrac{(11)^{1/2}}{2}t\right)$; $e^{t/2}(\cos 2t + \frac{1}{4}\sin 2t)$

1.14. $\frac{3}{10}\sinh \frac{5}{3}t$

1.16. $s^2 F(s) + 2sF(s) - F(s) - 1$

1.17. $2s^2 F(s) - 3sF(s) + F(s) - 4s + 12$

187

Section 1.6

1.20. $e^t + 2e^{-t} - 3e^{2t}$

1.21. $e^t(1 + 2t + t^2)$

1.22. $e^{-t}(\frac{1}{2}t^2 - \frac{1}{6}t^3)$

1.23. $2e^{3t} + e^{-2t} + te^{-2t}$

1.24. $\frac{1}{2} + \frac{1}{4}e^{2t} + \frac{1}{4}e^{-2t} = (\cosh t)^2$

1.25. $\sinh at - \sin at$

1.26. $\frac{1}{2}e^{3t} \sin 2t$

1.27. $at - \sin at$

1.28. $e^{-3t} - \sin 2t$

1.29. $2e^{t/2} \cos \dfrac{(11)^{1/2}}{2} t - \dfrac{2}{(11)^{1/2}} e^{t/2} \sin \dfrac{(11)^{1/2}}{2} t + e^{-t}$

1.30. $t \cos at$

1.31. $3e^t + \dfrac{9(2)^{1/2}}{8} \sin 2^{1/2}t - \dfrac{2^{1/2}}{2} t \sin 2^{1/2}t - \dfrac{1}{4} t \cos 2^{1/2}t$

Section 1.7

1.32. $x(t) = 2e^{-t}$

1.33. $x(t) = e^t - e^{-2t} - 3te^{-2t}$

1.34. $x(t) = (1 + t)e^{-2t}$

1.35. $x(t) = t + \cos t$

1.36. $x(t) = \cos t + \frac{1}{2}(\sin t - t \cos t)$

1.37. $x(t) = \frac{1}{8}e^{-t} \sin t + \frac{7}{24}e^{-t} \sin 3t$

1.38. $x(t) = \frac{1}{25}[6 \sin t - 8 \cos t + e^{-2t}(8 + 35t)]$

1.39. $x(t) = -te^t + \frac{1}{2}t^2e^t + \frac{1}{24}t^4e^t$

1.40. $x(t) = 5e^{-t} - 2e^{4t}$
$\quad\;\; y(t) = 5e^{-t} + 3e^{4t}$

1.41. $x(t) = -4e^t + 5e^{2t} - 1$
$\quad\;\; y(t) = -2e^t - e^{2t} + 3$

1.42. $x(t) = 1 - t - 3e^{-2t} - \cos t$
$\quad\;\; y(t) = t + 2 + e^{-2t} + \sin t$

1.43. $x(t) = \frac{3}{2} \sin t - \frac{1}{6} \sin 3t$
$\quad\;\; y(t) = \frac{3}{2} \sin t + \frac{1}{2} \sin 3t$

Section 1.8

1.44. $x(t) = 2 \cos 2t$

1.45. $x(t) = 2e^{-2t}(1 + 2t)$

1.46. $v(t) = 320 - 320e^{-t/10}$

1.47. $x(t) = 2 \cos 4t$

1.48. $x(t) = e^{-t/4}\left[\dfrac{2}{(255)^{1/2}} \sin \dfrac{(255)^{1/2}}{4} t + 2 \cos \dfrac{(255)^{1/2}}{4} t \right]$

1.49. $x(t) = 2 \cos 4t + \frac{1}{32}(\sin 4t - 4t \cos 4t)$

1.50. $q(t) = 6 - 6e^{-4t} \cos 3t - 8e^{-4t} \sin 3t$

1.52. $i(t) = \dfrac{E}{R}(1 - e^{-Rt/L})$

1.53. $i_1(t) = 6 - 6e^{-55t/2}; \quad i_2(t) = 2 - 2e^{-55t/2}$

1.54. $i_2(t) = (50)^{1/2}(e^{-[10-(50)^{1/2}]t} - e^{-[10+(50)^{1/2}]t})$

Section 2.2

2.1. $u(t) - u(t - \pi)$

2.2. $2[u(t - 1) - u(t - 3)]$

2.3. $u(t - 2\pi) \cos t$

2.4. $[u(t - 2) - u(t - 3)]t^2$

2.5. $\dfrac{6e^{-3s}}{s^4}$

2.6. (b) $\dfrac{se^{-2\pi s}}{s^2 + 4}$

2.7. $\dfrac{4e^{-3s}}{s^3} + \dfrac{15e^{-3s}}{s^2} + \dfrac{26e^{-3s}}{s}$

2.8. $L\left\{u(t - 2)\left[53 + 93(t - 2) + \dfrac{130}{2}(t - 2)^2 + \dfrac{132}{3!}(t - 2)^3\right.\right.$

$$\left.\left. + \dfrac{72}{4!}(t - 2)^4\right]\right\} = e^{-2s}\left(\dfrac{53}{s} + \dfrac{93}{s^2} + \dfrac{130}{s^3} + \dfrac{132}{s^4} + \dfrac{72}{s^5}\right)$$

2.10. $u(t - 5)(t - 5)$

2.11. $u(t) - u(t - 3)$

2.12. $\frac{1}{2}[u(t)t^2 - 2u(t - 1)(t - 1)^2 + u(t - 2)(t - 2)^2]$

2.13. $u(t - \pi) \cos 2t$

2.14. $u(t - a) \sinh(t - a)$

Section 2.3

2.15. (a) $\dfrac{1}{s^2} - \dfrac{e^{-2s}}{s^2} - \dfrac{2e^{-3s}}{s}$; (b) $\dfrac{e^{-s}}{s} - \dfrac{e^{-s}}{s^2} + \dfrac{e^{-2s}}{s^2}$

(c) $\dfrac{1}{s}(1 - 2e^{-as} + e^{-2as}) = \dfrac{(1 - e^{-as})^2}{s}$

(d) $2\left(\dfrac{1}{s^2} - \dfrac{e^{-s}}{s^2} - \dfrac{e^{-s}}{s} + \dfrac{e^{-2s}}{s^2} - \dfrac{e^{-3s}}{s^2} - \dfrac{e^{-3s}}{s}\right)$

(e) $\dfrac{(1 - e^{-s})(1 - e^{-2s})}{s^2}$

(f) $\dfrac{1}{a}\left(\dfrac{a}{s} - \dfrac{1}{s^2} + \dfrac{2e^{-as}}{s^2} - \dfrac{e^{-2as}}{s^2} - \dfrac{ae^{-2as}}{s}\right)$

Section 2.5

2.16. $x(t) = b \cos \dfrac{kt}{m^{1/2}} + \dfrac{F_0}{k^2} u(t - a)\left[1 - \cos \dfrac{k}{m^{1/2}}(t - a)\right]$

2.17. $q(t) = t - \dfrac{1}{(50)^{1/2}} \sin(50)^{1/2}t - u(t - 1)$

$$\times \left[t - \dfrac{1}{(50)^{1/2}} \sin(50)^{1/2}(t - 1) - \cos(50)^{1/2}(t - 1)\right]$$

2.18. (a) $\quad i(t) = \dfrac{E_0}{R}(1 - e^{-Rt/L})$

(b) $\quad i(t) = \begin{cases} \dfrac{E_0}{R}(1 - e^{-Rt/L}), & 0 \le t < a \\[2mm] -\dfrac{E_0}{R} e^{-Rt/L}(1 - e^{Ra/L}), & t \ge a \end{cases}$

or

$$i(t) = \dfrac{E_0}{R}[1 - e^{-Rt/L} - u(t - a)(1 - e^{-R(t-a)/L})]$$

2.19. $x(t) = \dfrac{1}{k^2}\left[1 - \cos kt - u\left(t - \dfrac{\pi}{k}\right)(1 + \cos kt)\right]$

$$= \begin{cases} \dfrac{1}{k^2}(1 - \cos kt), & 0 \le t < \dfrac{\pi}{k} \\[2mm] -\dfrac{2}{k^2} \cos kt, & t \ge \dfrac{\pi}{k} \end{cases}$$

2.20. $x(t) = \dfrac{1}{k^2}\left[X_1 k \sin kt - \cos kt + 1 - u\left(t - \dfrac{\pi}{k}\right)(\cos kt + 1)\right]$

2.21. $x(t) = \dfrac{1}{k^2 - 1}\, u(t)\left(\sin t - \dfrac{1}{k}\sin kt\right)$

$\qquad\qquad + \dfrac{1}{k^2 - 1}\, u(t - \pi)\left[\sin(t - \pi) - \dfrac{1}{k}\sin k(t - \pi)\right]$

$\qquad\quad = \begin{cases} \dfrac{1}{1 - k^2}\left(\dfrac{1}{k}\sin kt - \sin t\right), & 0 \le t < \pi \\[3mm] \dfrac{1}{k(1 - k^2)}\left[\sin kt + \sin k(t - \pi)\right], & t \ge \pi \end{cases}$

2.22. $x(t) = \begin{cases} \dfrac{1}{2k^2}\left(\sin kt - kt \cos kt\right), & 0 \le t < \dfrac{\pi}{k} \\[3mm] -\dfrac{\pi}{2k^2}\cos kt, & t \ge \dfrac{\pi}{k} \end{cases}$

2.23. $i(t) = \begin{cases} \dfrac{1}{R^2 + L^2\omega^2}\left(L\omega e^{-Rt/L} - L\omega \cos \omega t + R\sin \omega t\right), & 0 \le t < \pi/\omega \\[3mm] \dfrac{1}{R^2 + L^2\omega^2}\left[L\omega e^{-Rt/L}(1 + e^{R\pi/L\omega})\right], & t \ge \pi/\omega \end{cases}$

2.24. (a) $x(t) = X_0 \cos kt + \dfrac{t}{k^2} - \dfrac{1}{k^3}\sin kt$

(b) $x(t) = \begin{cases} X_0 \cos kt + \dfrac{t}{k^2} - \dfrac{1}{k^3}\sin kt, & 0 \le t < \pi \\[3mm] X_0 \cos kt - \dfrac{1}{k^3}\sin kt \pm \dfrac{1}{k^3}\sin kt \pm \dfrac{\pi}{k^2}\cos kt, & t \ge \pi \end{cases}$

2.25. $Y(x) = \dfrac{F_0 a^2 x^2}{4EI} - \dfrac{F_0 a x^3}{6EI} + \dfrac{F_0 x^4}{24EI}$

2.26. $Y(x) = \dfrac{F_0 x^2(a - x)^2}{24EI}$

2.27. $Y(x) = \dfrac{F_0}{EI}\left[\dfrac{a^2 x^2}{16} - \dfrac{a x^3}{12} + \dfrac{x^4}{24} - \dfrac{1}{24}u\left(x - \dfrac{a}{2}\right)\left(x - \dfrac{a}{2}\right)^4\right]$

2.28. $Y(x) = \dfrac{F_0}{EI}\left[\dfrac{a^3 x^2}{48} - \dfrac{a^2 x^3}{48} + \dfrac{x^5}{5!} - u\left(x - \dfrac{a}{2}\right)\dfrac{(x - a/2)^5}{5!}\right.$

$\left. - \dfrac{a}{2}u\left(x - \dfrac{a}{2}\right)\dfrac{(x - a/2)^4}{4!}\right]$

2.29. $x(t) = b\cos\dfrac{k}{m^{1/2}}t + \dfrac{F_0}{km^{1/2}}u(t - a)\sin\dfrac{k}{m^{1/2}}(t - a)$

2.30. $x(t) = b\cos\dfrac{k}{m^{1/2}}t + \dfrac{A}{k^2 - m\omega^2}\sin\omega t - \dfrac{A\omega m^{1/2}}{k(k^2 - m\omega^2)}\sin\dfrac{k}{m^{1/2}}t$

$- \dfrac{2}{km^{1/2}}u(t - 10)\sin\dfrac{k}{m^{1/2}}(t - 10)$

2.31. $q(t) = \dfrac{20(11)^{1/2}}{33}[u(t - 1)e^{-10(t-1)}\sin 30(11)^{1/2}(t - 1)$

$+ u(t - 2)e^{-10(t-2)}\sin 3(11)^{1/2}(t - 2)]$

2.32. $q(t) = E_0\left(\dfrac{C}{L}\right)^{1/2}\sin\dfrac{t}{(LC)^{1/2}}$

2.33. $q(t) = \dfrac{E_0}{\alpha L}e^{-Rt/2L}\sin\alpha t,\quad$ where $\alpha = \left(\dfrac{1}{LC} - \dfrac{R^2}{4L^2}\right)^{1/2}$

2.34. $Y(x) = \dfrac{P_0}{EI}\left[\dfrac{2ax^2}{27} - \dfrac{10x^3}{81} + \dfrac{1}{6}u\left(x - \dfrac{a}{3}\right)\left(x - \dfrac{a}{3}\right)^3\right]$

2.35. $Y(x) = \dfrac{a}{16}\left(\dfrac{P_1}{EI} + \dfrac{3P_2}{8EI}\right)x^2 - \dfrac{1}{12}\left(\dfrac{P_1}{EI} + \dfrac{5P_2}{16EI}\right)x^3$

$+ \dfrac{P_1}{6EI}u\left(x - \dfrac{a}{2}\right)\left(x - \dfrac{a}{2}\right)^3 + \dfrac{P_2}{6EI}u\left(x - \dfrac{3a}{4}\right)\left(x - \dfrac{3a}{4}\right)^3$

2.36. $Y(x) = \dfrac{F_0 x^2}{48EI}(2x^2 - 4ax + 3a^2) + \dfrac{P_0 x^2}{24EI}(9a - 4x)$

$- \dfrac{F_0}{24EI}u\left(x - \dfrac{a}{2}\right)\left(x - \dfrac{a}{2}\right)^4 + \dfrac{P_0}{6EI}u\left(x - \dfrac{3a}{4}\right)\left(x - \dfrac{3a}{4}\right)^3$

2.37. $Y(x) = \dfrac{P_0 x^2}{6EI}(3a - x)$

2.38. $Y(x) = \dfrac{P_0}{EI}\left[\dfrac{ax^2}{4} - \dfrac{x^3}{6} + \dfrac{1}{6}\,u\left(x - \dfrac{a}{2}\right)\left(x - \dfrac{a}{2}\right)^3\right]$

2.39. $Y(x) = \dfrac{P_0(5a^2 - 9x^2)x}{81EI} + \dfrac{P_0}{6EI}\,u\left(x - \dfrac{a}{3}\right)\left(x - \dfrac{a}{3}\right)^3$

2.40. $x_1(t)$ contains exponential decaying terms, indicating that the mass m_1 has its motion damped out (*vibration absorber*), for if $-a + bi$ is a root, $s + a - bi$ is a factor, and $x_1(t)$ contains the term e^{-at}.

Section 2.6

2.43. (a) $\dfrac{1}{s(1 + e^{-as})}$

(b) $\dfrac{1}{s} - \dfrac{e^{-as}}{s} + \dfrac{e^{-2as}}{s} - \dfrac{e^{-3as}}{s} + \cdots$

2.44. (a) $\dfrac{1}{s^2} - \dfrac{e^{-s}}{s(1 - e^{-s})}$

(b) $\dfrac{1}{s^2} - \dfrac{1}{s}\,(e^{-s} + e^{-2s} + e^{-3s} + \cdots)$

2.45. (a) $\dfrac{\omega}{s^2 + \omega^2} \cdot \dfrac{1}{1 - e^{-\pi\omega/s}}$

(b) $\dfrac{\omega}{s^2 + \omega^2}\,(1 + e^{-\pi s/\omega} + e^{-2\pi s/\omega} + e^{-3\pi s/\omega} + \cdots)$

2.46. (a) $\dfrac{\omega}{s^2 + \omega^2}\coth\dfrac{\pi s}{2\omega} = \dfrac{\omega(1 + e^{-\pi s/\omega})}{(s^2 + \omega^2)(1 - e^{-\pi s/\omega})}$

(b) $\dfrac{\omega}{s^2 + \omega^2} + \dfrac{2\omega}{s^2 + \omega^2}\,(e^{-\pi s/\omega} + e^{-2\pi s/\omega} + e^{-3\pi s/\omega} + \cdots)$

Section 2.7

2.47. $v(t) = \dfrac{P}{m}\,[u(t) + u(t - a) + u(t - 2a) + \cdots]$

2.48. $q(t) = E_0\left(\dfrac{C}{L}\right)^{1/2}\left[u(t)\sin\dfrac{t}{(LC)^{1/2}} + u(t - a)\sin\dfrac{t - a}{(LC)^{1/2}} + \cdots\right]$

2.49. $q(t) = \dfrac{20(11)^{1/2}}{33} \, [u(t)e^{-10t} \sin 30(11)^{1/2}t$

$\qquad\qquad + \, u(t-1)e^{-10(t-1)} \sin 30(11)^{1/2}(t-1) + \cdots]$

2.50. (a) $i(t) = \dfrac{E_0}{L} \, e^{-Rt/L}(1 + e^{Ra/L} + e^{2Ra/L} + \cdots + e^{nRa/L})$

$\qquad\qquad = \dfrac{E_0 e^{-Rt/L}(1 - e^{(n+1)Ra/L})}{L(1 - e^{Ra/L})},$

$\qquad\qquad\qquad\qquad\qquad na \le t < (n+1)a, \; n = 0, 1, 2, \ldots$

$\qquad$ (b) $i(t) = \dfrac{E_0 e^{-R(t-na)/L}}{L(1 - e^{-Ra/L})}$

2.51. $i(t) = e^{-Rt/L}\left[I_0 + \dfrac{E_0(1 - e^{nRa/L})}{L(1 - e^{Ra/L})}\right], \qquad (n-1)a \le t < na$

2.52. $q(t) = \dfrac{E_0}{L\beta} \, [u(t)e^{-\alpha t} \sin \beta t + u(t-a)e^{-\alpha(t-a)} \sin \beta(t-a) + \cdots],$

$\qquad\qquad\qquad$ where $\alpha = \dfrac{R}{2L}$ and $\beta = \left(\dfrac{1}{LC} - \dfrac{R^2}{4L^2}\right)^{1/2}$

2.53. (a) $i(t) = \begin{cases} \dfrac{E_0}{R} - \dfrac{E_0}{R} \, e^{-Rt/L}\left(\dfrac{1 + e^{(n+1)Ra/L}}{1 + e^{Ra/L}}\right), & n \text{ even} \\[4mm] -\dfrac{E_0}{R} \, e^{-Rt/L}\left(\dfrac{1 - e^{(n+1)Ra/L}}{1 + e^{Ra/L}}\right), & n \text{ odd,} \end{cases}$

$\qquad\qquad\qquad\qquad\qquad na \le t < (n+1)a, \; n = 0, 1, 2, \ldots$

$\qquad$ (b) $i(t) = \begin{cases} \dfrac{E_0}{R} - \dfrac{E_0}{R} \, \dfrac{e^{-R(t-na)/L}}{e^{-Ra/L} + 1}, & n \text{ even} \\[4mm] \dfrac{E_0}{R} \, \dfrac{e^{-R(t-na)/L}}{e^{-Ra/L} + 1}, & n \text{ odd} \end{cases}$

$\qquad$ (c) $L\dfrac{di}{dt} = \begin{cases} \dfrac{E_0 e^{-R(t-na)/L}}{e^{-Ra/L} + 1}, & n \text{ even} \\[4mm] \dfrac{-E_0 e^{-R(t-na)/L}}{e^{-Ra/L} + 1}, & n \text{ odd} \end{cases}$

2.54. $i(t) = -\dfrac{E_0}{R} + \dfrac{E_0}{R} e^{-Rt/L} + 2A(t)$, where $A(t)$ is the solution to

Exercise 2.53(a)

2.55. $x(t) = -(2n + 1)\dfrac{P}{k^2} \cos kt + K$, where $K = \begin{cases} \dfrac{P}{k^2}, & n \text{ even} \\[2ex] -\dfrac{P}{k^2}, & n \text{ odd,} \end{cases}$

$$n\frac{\pi}{k} \le t < (n + 1)\frac{\pi}{k}, \ n = 0, 1, \ldots$$

2.56. $x(t) = \dfrac{X_1}{k} \sin kt + B(t)$, where $B(t)$ is the solution to Exercise 2.55

2.57. $i(t) = -\dfrac{Le^{-Rt/L}}{R^2 + L^2} + \dfrac{2Le^{-Rt/L}(1 - e^{(n+1)R\pi/L})}{(R^2 + L^2)(1 - e^{R\pi/L})} \pm \dfrac{L}{R^2 + L^2} \cos t$

$\mp \dfrac{R}{R^2 + L^2} \sin t, \qquad n\pi \le t < (n + 1)\pi, \ n = 0, 1, 2, \ldots$

2.58. $x(t) = X_0 \cos kt + \dfrac{t}{k^2} - \dfrac{1}{k^3} \sin kt$

$- \dfrac{\pi}{k^2} \displaystyle\sum_{n=1}^{\infty} u(t - n\pi)\, [1 - \cos k(t - n\pi)]$

Section 5.1

5.1. (a) $1 - \cos at$;   (b) $t \sin at$;   (c) $\sin at + at \cos at$

5.3. $x(t) = ae^t$

5.4. $x(t) = \sinh t$

5.5. $x(t) = c$

5.6. $x(t) = a^{1/2}$; $x(t) = -a^{1/2}$

5.7. $x(t) = \pm a^{1/2} J_0(t)$

5.8. $i(t) = 5000te^{-400t}$

Section 5.3

5.11. $\dfrac{X_0 c_1 k_2 - F_0 m_2}{m_1 m_2}$

5.12. $i = \dfrac{dq}{dt}$

5.13. $e_2(t) = E_1 e^{-t/RC} - E_1 u(t - a) e^{-(t-a)/RC}$

5.15. $i(t) = \dfrac{1}{R}\,\delta(t) - \dfrac{1}{R^2 C}\,e^{-t/RC}$

5.16. $Ri(t) = \begin{cases} \dfrac{E_0 e^{-(t-na)/RC}}{e^{-a/RC} + 1}, & n \text{ even} \\[4mm] \dfrac{-E_0 e^{-(t-na)/RC}}{e^{-a/RC} + 1}, & n \text{ odd}, \end{cases}$

$$an \le t < (n+1)a, \quad n = 0, 1, 2, \ldots$$

5.17. $i(t) = \begin{cases} \dfrac{1}{1 + R^2 C^2 \omega^2}\left(\dfrac{1}{R}\,e^{-t/RC} + RC^2\omega^2 \cos \omega t - C\omega \sin \omega t\right), \\[4mm] \hspace{6cm} 0 \le t < \dfrac{\pi}{\omega} \\[4mm] \dfrac{1}{1 + R^2 C^2 \omega^2}\,\dfrac{1}{R}\,e^{-t/RC}(1 + e^{\pi/RC}), \hspace{1cm} t \ge \dfrac{\pi}{\omega} \end{cases}$

5.18. $i(t) = -\dfrac{E_0}{R}\,e^{-t/RC} + \dfrac{2E_0}{R}\,e^{-t/RC}\left[\dfrac{1 - (-e^{a/RC})^{n+1}}{1 + e^{a/RC}}\right],$

$$na \le t < (n+1)a, \quad n = 0, 1, 2, \ldots$$

5.19. $i(t) = C - Ce^{-t/RC} - \dfrac{\pi}{R}\displaystyle\sum_{n=1}^{\infty} u(t - n\pi)e^{-(t-n\pi)/RC}$

5.20. $i(t) = \dfrac{1}{R[\omega^2 + (1/R^2C^2)]}\left\{\dfrac{1}{RC}\cos \omega t + \omega \sin \omega t - \dfrac{1}{RC}\,e^{-t/RC}\right.$

$$+ 2\sum_{n=1}^{\infty} u\left(t - \frac{n\pi}{\omega}\right)\left[\frac{1}{RC}\cos \omega\left(t - \frac{n\pi}{\omega}\right) + \omega \sin \omega\left(t - \frac{n\pi}{\omega}\right)\right.$$

$$\left.\left. - \frac{1}{RC}\,e^{-(1/RC)[t-(n\pi/\omega)]}\right]\right\}$$

Section 7.3

7.1. $U(x, t) = U_1 + (U_0 - U_1) \operatorname{erfc}\left[\dfrac{x}{2(kt)^{1/2}}\right]$

7.2. $U(x, t) = U_0 \operatorname{erfc}\left[\dfrac{x}{2(kt)^{1/2}}\right]$

7.3. $U(x, t) = \dfrac{Q_0(x - 1)}{2(\pi k t^3)^{1/2}}\, e^{-(x-1)^2/4kt}$

7.4. $U(x, t) = \displaystyle\int_0^t F(t - u)\,\dfrac{e^{-x^2/4u}}{(\pi u)^{1/2}}\, du$

7.5.   (a) $U(x, t) = K \displaystyle\int_0^t \dfrac{e^{-x^2/4u}}{(\pi u)^{1/2}}\, du;$ (b) $U(x, t) = \dfrac{K}{(\pi t)^{1/2}}\, e^{-x^2/4t}$

7.6. $V(x, t) = V_0 \operatorname{erfc}\left[\dfrac{x}{2}\left(\dfrac{RC}{t}\right)^{1/2}\right]$

7.7. $Y(x, t) = u\left(t - \dfrac{x}{a}\right)G\left(t - \dfrac{x}{a}\right)$

7.8. $Y(x, t) = \cos\dfrac{a\pi t}{l}\, \sin\dfrac{\pi x}{l}$

7.9. $Y(x, t) = \left[u\left(t - \dfrac{x}{a}\right) - u\left(t - \dfrac{x}{a} - 1\right)\right]\left(t - \dfrac{x}{a}\right)$

7.10. $Y(x, t) = \delta\left(t - \dfrac{x}{a}\right)$

7.11. $U(x, t) = \begin{cases} K \operatorname{erfc}\left[\dfrac{x}{2(kt)^{1/2}}\right], & 0 \le t < 1 \\[2ex] K \operatorname{erfc}\left[\dfrac{x}{2(kt)^{1/2}}\right] - K \displaystyle\int_1^t \dfrac{x}{2(\pi k u^3)^{1/2}}\, e^{-x^2/4ku}\, du, & t \ge 1 \end{cases}$

Section 7.4

7.12. $U(x, t) = \dfrac{4l}{\pi} \displaystyle\sum_{n=1}^{\infty} \dfrac{(-1)^{n+1}}{n} e^{-n^2\pi^2 t/l^2} \sin \dfrac{n\pi x}{l}$

7.13. $U(x, t) = \dfrac{4K}{\pi} \displaystyle\sum_{n=1}^{\infty} \dfrac{1}{2n-1} e^{-(2n-1)^2\pi^2 t} \sin(2n-1)\pi x$

7.14. $U(x, t) = 2x + \dfrac{2}{\pi} \displaystyle\sum_{n=1}^{\infty} \dfrac{(-1)^n}{n} [2 - K + (-1)^n K] e^{-n^2\pi^2 t} \sin n\pi x$

7.15. $U(x, t) = U_1$

$$+ \frac{4(U_1 - U_0)}{\pi} \sum_{n=1}^{\infty} \frac{(-1)^n}{2n-1} e^{-(2n-1)^2\pi^2 kt/4l^2} \cos \frac{(2n-1)\pi x}{2l}$$

7.16. $Y(x, t) = \dfrac{\sin(\pi x/al) \sin(\pi t/l)}{\sin(\pi/a)} + \dfrac{2a}{\pi} \displaystyle\sum_{n=1}^{\infty} \dfrac{(-1)^n}{a^2 n^2 - 1} \sin \dfrac{n\pi x}{l} \sin \dfrac{n a\pi t}{l}$

7.17. $Y(x, t) = \dfrac{gx}{2a^2} (2l - x)$

$$- \frac{16gl^2}{\pi^3 a^2} \sum_{n=1}^{\infty} \frac{1}{(2n-1)^3} \sin \frac{(2n-1)\pi x}{2l} \cos \frac{(2n-1)\pi at}{2l}$$

7.18. $U(x, t) = -\dfrac{b}{K}\left[x - \dfrac{8}{\pi^2} \displaystyle\sum_{n=1}^{\infty} \dfrac{(-1)^{n-1}}{(2n-1)^2} e^{-(2n-1)\pi^2 t/4} \sin \dfrac{(2n-1)\pi x}{2}\right]$

(Heat is extracted at a constant rate.)

7.19. $U(x, t) = K\left[\dfrac{x^3 - x}{6} + xt + \dfrac{2}{\pi^3} \displaystyle\sum_{n=1}^{\infty} \dfrac{(-1)^{n-1}}{n^3} e^{-n^2\pi^2 t} \sin n\pi x\right]$

7.20. $V(x, t) = \dfrac{F_0}{E}\left[x + \dfrac{8l}{\pi^2} \displaystyle\sum_{n=1}^{\infty} \dfrac{(-1)^n}{(2n-1)^2} \sin \dfrac{(2n-1)\pi x}{2l} \cos \dfrac{(2n-1)\pi at}{2l}\right]$

Index